T0073630

BEAST COMPANIONS

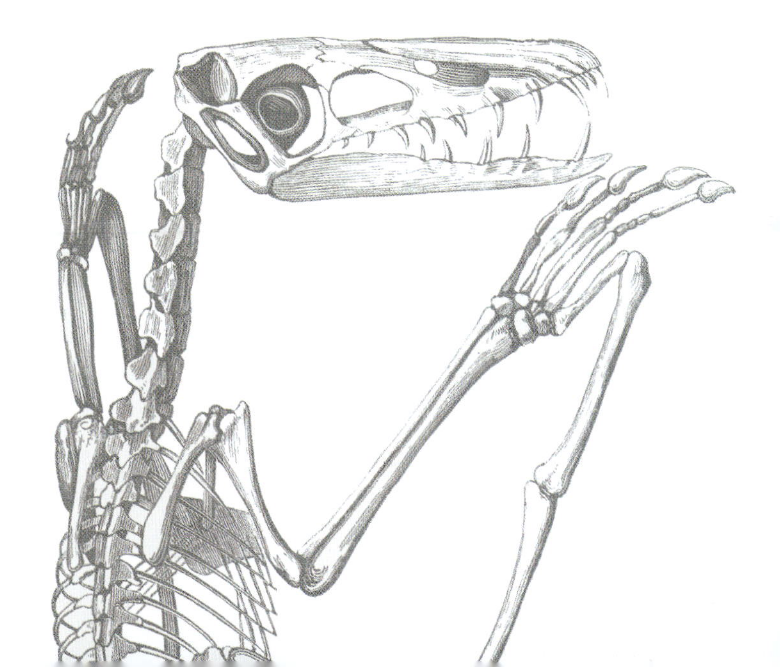

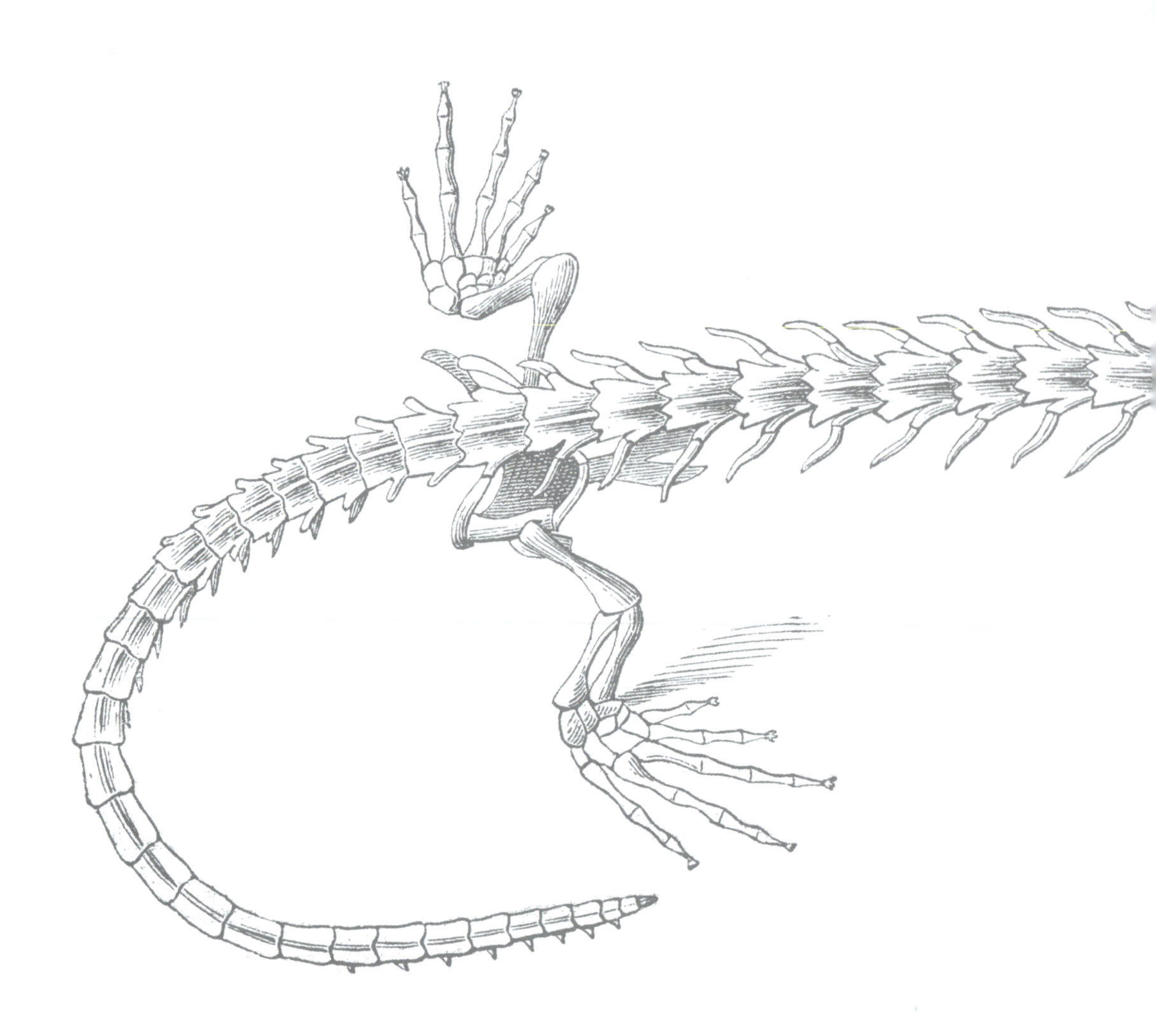

Life of the Past Thomas Holtz, editor

JOHN FOSTER

BEAST COMPANIONS

THE UNSUNG ANIMALS OF THE DINOSAURS' WORLD

Indiana University Press

This book is a publication of

Indiana University Press
Office of Scholarly Publishing
Herman B Wells Library 350
1320 East 10th Street
Bloomington, Indiana 47405 USA

iupress.org

This book is printed on acid-free paper.

Manufactured in China

First printing 2024

Library of Congress
Cataloging-in-Publication Data

Names: Foster, John Russell, author.
Title: Beast companions : the unsung animals of the dinosaurs' world / John Foster.
Description: Bloomington, Indiana : Indiana University Press, [2024] | Series: Life of the past | Includes bibliographical references.
Identifiers: LCCN 2023046230 (print) | LCCN 2023046231 (ebook) | ISBN 9780253069405 (hardback) | ISBN 9780253069412 (ebook)
Subjects: LCSH: Paleontology—Mesozoic. | Animals, Fossil. | Evolution (Biology) | BISAC: SCIENCE / Paleontology | NATURE / Animals / Dinosaurs & Prehistoric Creatures
Classification: LCC QE731 .F67 2024 (print) | LCC QE731 (ebook) | DDC 560/.176—dc23/eng/20240126
LC record available at https://lccn.loc.gov/2023046230
LC ebook record available at https://lccn.loc.gov/2023046231

For

ReBecca, Ruby, and Harrison

And in memory of a few friends-mentors:

Dan Varner, Fred Peterson, Richard Stucky, and Martin Lockley

The most impressive aspect of the living world is its diversity. . . .
Wherever one looks in nature, one finds uniqueness.

Ernst Mayr, *This Is Biology*, 1997

Yet for all this exuberance and flair there are constraints:
convergence is inevitable, yet paradoxically the net result
is not one of sterile returns to worn-out themes; rather
there is also a patent trend of increased complexity.

Simon Conway Morris, *Life's Solution*, 2003

Contents

Preface

They fascinate nearly everyone at some point, and in some cases, for lifetimes. They were among the most impressive vertebrates in Earth's history, reaching a known diversity of thousands of species and dominating terrestrial ecosystems for 160 million years. Some dinosaurs are household names even in homes with neither paleontologists nor six-year-olds under the roof. *Tyrannosaurus*, *Argentinosaurus*, *Velociraptor*, *Stegosaurus*. The names and profiles are known to many, even if the details are not. To others, the details are known and frequently debated. Dinosaurs are so famous they have become, on occasion, incorrectly synonymized with "fossil" of any kind. And they were so successful for so long that the colloquial pejorative "dinosaur" of modern, nonbiological usage quietly undermines most people's intended meaning in its use. Don't be a dinosaur? Don't diversify into an amazing array of forms, adapt to a wide range of environments all over the world, and last as a group for 16,000 times longer than human civilization so far? We should all be so lucky.

The nonavian dinosaurs serve as the gold standard of paleontology as examples of adaptation and diversification and ecosystem dominance. This standard comes from their obvious utility, however, and not necessarily from them being the best examples of everything we learn from them. In many cases, they are simply the flashiest examples. They may not be the most important of fossils, but they are (in many cases) the biggest and most recognizable, and they demonstrate aspects of Earth's biological history in large, bold, red caps perhaps more so than other groups.

Yet for all their fame and reputation, for all we can learn from them about evolution, extinction, and the Earth in general in deep time, dinosaurs are only half the story during the Mesozoic era, also known as the Age of Dinosaurs. Many other, often equally remarkable, animals lived during that time, over, under, and around the dinosaurs, ancient species from which we can also learn about the biotic history of our planet. And in some cases, these animals reveal just as much and often *more* about the extinct ecosystems of the time of dinosaurs as the dinosaurs themselves.

Our journey here will be to visit a few of the groups of animals that also occupied the environments of the Mesozoic between 230 and 66 million years ago, alongside the dinosaurs. Many are groups still familiar to us today, and some are rather different or at least contained members that are very different from what we see around us in the world today. If we are all familiar with the dinosaurs that lived from the Triassic through the Jurassic periods and to the end of the Cretaceous period, I hope that by the end of this book, you will feel more familiar with the *world* of the dinosaurs, with what made it unique in its mix of animals, and in what ways it shared common elements with the world of today.

The format for our journey will include an introduction to the Mesozoic in general and its early millennia (chap. 1); then we will visit each group by chapter. Although we will concentrate on the vertebrate cousins of the dinosaurs, and although it is tempting to visit the climates and vegetation of each individual time period, we must skip these for space (another book of its own) and focus on the animals. The first group chapter (chap. 2) will tour several of the invertebrate groups known from this time period, then check in with the fish of the age (chap. 3), which had been around since the early Paleozoic, well before dinosaurs. After that, we'll meet various vertebrate groups familiar and foreign before a concluding chapter on the fates of the nondinosaurs since the Cretaceous (chap. 14). In between, we will meet the first frogs and salamanders (chap. 4), turtles (chap. 5), rhynchocephalians (chap. 6), squamates (chap. 7), marine reptiles (chap. 8), choristoderes (chap. 9), crocodiles (chap. 10), pterosaurs (chap. 11), the avian dinosaurs we know as birds (chap. 12), and mammals (chap. 13) of the Mesozoic.

You'll undoubtedly notice in this journey that my personal experience with these groups comes largely from North America, especially the Rocky Mountain region in the western United States. And my specialty (if you can call it that), which will again be fairly obvious, is in the Late Jurassic. But the nondinosaurian groups are found all around the globe, and the most important sites for individual groups are often in places farther afield than my geological backyard. In fact, most are on other continents, largely Asia, Europe, and South America. I have included these to the best of my ability, despite having visited very few of them myself.

This is not intended to be an exhaustive review of all nondinosaurian groups (plus birds) of the Mesozoic; there are too many to cover here. Rather, I hope to give the reader a taste of how diverse the Mesozoic was, beyond the well-known cast of dinosaurs. Think of it as highlights and an invitation to delve deeper into the equally interesting supporting cast of dinosaur times.

John Foster
Vernal, Utah
August 9, 2022

Acknowledgments

Although the broad animal groups featured in this book are nearly all ones I've generally encountered in my fieldwork or lab work at some point or another, none of them is exactly what I could consider a taxonomic specialty of mine. Input, feedback, and honest opinions from many of those actual specialists in the various groups featured here have been invaluable, both during this project and over the years of wrestling paleontological topics. Especially helpful have been Matt Bonnan, Adam Huttenlocker, Tom Holtz, Andy Heckert, Randy Nydam, Jaelyn Eberle, Andrew Milner, Spencer Lucas, Marc Jones, Thomas Adams, Kirk Johnson, Ash Poust, Bhart-Anjan Bhullar, Ryoko Matsumoto, Bruce Schumacher, and Tom Stidham, each of whom graciously read parts of chapters (or the entire thing!) and helped improve the accuracy and currency of much of the data.

Other mentors and colleagues over the years have also proven critical in training me in various field and lab techniques, providing me with or alerting me to publications, answering random questions, teaming up in the field and museums, and otherwise influencing my thinking regarding the mostly nondinosaurian animals of the Mesozoic and their world. All these folks have been critical in ways large or small (and thus, cumulatively, immensely), and I am indebted to each. In addition to all the folks mentioned above, these include (in no particular order): Scott Madsen, Dave Archibald, Randy Irmis, Susan Evans, Neil Kelley, Jim Clark, Don Prothero, Jim Martin, Luis Chiappe, Gabe Bever, Darrin Pagnac, Kelli Trujillo, Alan Titus, Brian Davis, Guillermo Rougier, Jason Lillegraven, George Callison, Janet Gillette, Martin Lockley, Bill Turnbull, Jim Kirkland, Mike Greenwald, Phil Bjork, George Engelmann, John Damuth, Zhexi Luo, Ken Carpenter, Dan Chure, and Peter Robinson.

Special thanks to the many research and collections colleagues at institutions around the world who helped provide images for the figures here (in very loose chronological order): Jeff Martz, Nick Fraser, Andy Heckert, Maria Lara, Dionizio Moura-Júnior, Tom Howells, Kevin Bylund, Daniel Blake, Mike Everhart, Dan Varner, Dave Archibald, Gloria Bader, Jim Kirkland, Martin Ebert, Christina Ifrim, Alexander Nützel, Barbara Grandstaff, Sanaa El-Sayed, Allison Tumarkin-Deratzian, Hesham Sallam, Matt Lamanna, Liping Dong, Angela Delgado-Buscalioni, Bo Wang, Susan Evans, Márton Rabi, Walter Joyce, Mark Ryan, Vida Horvat, Brian Grace, Spencer Lucas, Alex Hastings, Charlie Iverson, Erin Gredell, Marc Jones, Randy Nydam, Dean Lomax, Emily Berry, Darrin Pagnac, Mike Ray, Ryoko Matsumoto, ReBecca Hunt-Foster, Wei Gao, Jim Clark, Emily Berry, Mike Ray, Randy Irmis, Rob Gaston, Luis Chiappe, Stephanie

Abramowicz, Octavio Mateus, Guillermo Rougier, Zhexi Luo, Susie Maidment, Oliver Rauhut, Carrie Levitt-Bussian, Piergiulio Cappelletti, Bernhard Zipfel, Dave DeMar, Lance Grande, Rod Scheetz, Honggang Zhang, Oliver Wings, Torsten Scheyer, Jin Meng, Rachel Unruh, Laura Wilson, Annalisa Aiello, Julia McHugh, Andrey Sennikov, Laura Uglean-Jackson, James Hagadorn, Megan Blackwelder, Yves Laurent, and Georg Heumann.

Thanks to Gary Dunham, Tony Brewer, Brian Carroll, Anna Francis, Emma Getz, Brenna Hosman, Katie Huggins, David Hulsey, David Miller, Dan Pyle, Pamela Rude, Stephen Williams, and Vinodhini Kumarasamy for pulling this project together.

Finally, thanks once again to my family for tolerating my weekends at the downstairs work desk or with the laptop on the back deck—and for only sneaking up like ninjas occasionally.

Institutional Abbreviations

AMNH American Museum of Natural History, New York, New York, US

BDL Menat Museum, Bord du Lac, France

BMNH Beijing Museum of Natural History, Beijing, China

BP/I Evolutionary Studies Institute at the University of the Witwatersrand (Bernard Price Institute), Johannesburg, South Africa

BYU Brigham Young University Museum of Paleontology, Provo, Utah, US

CAGS Chinese Academy of Geological Sciences, Beijing, China

CM Carnegie Museum of Natural History, Pittsburgh, Pennsylvania, US

CMNH Cleveland Museum of Natural History, Cleveland, Ohio, US

DINO Dinosaur National Monument collections, Vernal, Utah, US

DMNH Denver Museum of Nature and Science, Denver, Colorado, US

FHPR Utah Field House of Natural History State Park Museum, Vernal, Utah, US

IVPP Institute of Vertebrate Paleontology and Paleoanthropology, Beijing, China

JME Jura Museum Eichstätt, Eichstätt, Germany

JZMP Jinzhou Museum of Paleontology, Jinzhou, China

LACM Natural History Museum of Los Angeles County, Los Angeles, California, US

MCCM Museo de Cuenca Castilla La Mancha, Cuenca, Spain

MCT Museu de Ciências da Terra, Companhia de Pesquisas de Recursos Minerais—Rio de Janeiro, Rio de Janeiro, Brazil

MCZ Museum of Comparative Zoology, Harvard University, Cambridge, Massachusetts, US

MHNT Muséum de Toulouse, Toulouse, France

MPCA Museo Provincial Carlos Ameghino, Cipolletti, Río Negro Province, Argentina

MPN Museu di Paleontologia, Napoli, Italy

MPSC Museu Placido Cidade Nuvens, Santana do Cariri, Brazil

MWC Museums of Western Colorado, Fruita, Colorado, US

NHMUK Natural History Museum, London, England

NIGP Nanjing Institute of Geology and Paleontology, Nanjing, China

NKMB Museum für Naturkunde, Humboldt-Universität, Berlin, Germany

NMMNH New Mexico Museum of Natural History and Science, Albuquerque, New Mexico, US

NMS National Museums of Scotland, Edinburgh, Scotland

PIMUZ Paleontological Institute and Museum, University of Zurich, Zurich, Switzerland

PIN RAS Paleontological Institute, Russian Academy of Sciences, Moscow, Russia

PMOL-SGP Paleontological Museum of Liaoning, Liaoning, China

PVL Instituto Miguel Lillo, Tucumán, Argentina

SBEI Shiramine Board of Education Ishikawa, Japan

SDSM South Dakota School of Mines and Technology, Museum of Geology, Rapid City, South Dakota, US

SMM Science Museum of Minnesota, St. Paul, Minnesota, US

SMMP Sanya Museum of Marine Paleontology, Hainan Province, China

SNSB-BSPG Staatliche Naturwissenschaftliche Sammlungen Bayerns, Bayerische Staatssammlung für Paläontologie und Geologie, Munich, Germany

STMN Shandong Tianyu Museum of Nature, Pingyi, China

TMP Royal Tyrrell Museum of Palaeontology, Drumheller, Alberta, Canada

UCM University of Colorado Museum, Boulder, Colorado, US

UMNH Natural History Museum of Utah, Salt Lake City, Utah, US

USNM National Museum of Natural History, Smithsonian Institution, Washington, DC, US

YPM Yale Peabody Museum, New Haven, Connecticut, US

BEAST COMPANIONS

Shadows in the Rain

Beginnings

1

AN OTHERWISE UNREMARKABLE piece of siltstone was the focus of my attention. Light gray in color and partly laminated, it also contained small green clay balls visible in cross section on its top surface. A few specks of carbonized plant remains and a tiny unidentifiable bone fragment or two were also visible but were not of interest. I had come to this pit in the Upper Jurassic Morrison Formation from another nearby site late that afternoon to pick up a bucketful of siltstone to check under the microscope back in the lab. Small bones less than 2.5 cm (1 in) long were the hoped-for prize. After a full day of chiseling under a giant *Camarasaurus* pelvis, this was a late break to change the pace, but the payoff was to come later. I hoped. Since the previous year, my crew had found a few fragments of mammal jaws and teeth in this pit in Wyoming, and I was hoping to find more.

As I wedged up another piece of siltstone with a dental pick, I was more intent on filling the bucket with matrix to check later than I was expecting to see much now as I worked. But then that one piece came up, flipping back to let late-afternoon July sunlight fall on a mammal jaw for the first time in 150 million years. I paused for a second in shock. The light glinted off the cusps of six tiny teeth preserved in position in a jaw just over an inch long, the teeth and bone shiny black with minerals emplaced during the millennia spent underground since the jaw's original burial. After a few moments, the initial jolt wore off, and I spent a few minutes lying on my stomach, eyes inches from the jaw, simply admiring the unexpected fossil of an early mammalian cousin that lived with *Allosaurus*, *Brontosaurus*, and *Stegosaurus*.

This was by far the best-preserved mammal we'd found in the quarry, and the next thing to ponder was how to get it out of the ground intact. It needed to be taken in a piece of siltstone about the size of a teacup saucer, but the trick would be to get it out without any cracks developing in unwanted areas, namely, through the jaw. I went back to the truck to get better tools and spent the next hour and a half removing the chunk with the jaw in it.

The jaw turned out to be that of a dryolestid mammal, a group that shared a common ancestor with that of modern mammals but not a group directly ancestral to our modern placental and marsupial mammals. Dryolestids were neither particularly rare nor small as Late Jurassic mammals went, but they do make gorgeous and impressive fossils, with very small teeth with a handful of acuminate cusps each and often as many as eight molars emplaced in a single lower jaw. The back of the lower jaw also still had postdentary bones not yet fully incorporated into the middle ear. We'll get to more of these trends in a later chapter. Dryolestids are a classic example of dinosaur contemporaries that were partly but not fully modern, ancient and primitive but clearly on their way to being familiar.

For much of my career in paleontology, I've worked at least part of the time on what are commonly known in the field as *microvertebrates*—bone-possessing chordates that are on the minute end of the size spectrum. As individuals, they certainly are not microscopic, and even their bones and teeth are not usually microscopic. Even though they are sometimes tiny, their remains can usually be seen with the unaided eye. But even the remains of larger microvertebrates are small enough that a microscope is almost always necessary to study them. In the age of the giant reptiles of the Mesozoic, the microvertebrates most likely encompassed just about all the vertebrates other than the dinosaurs (at least after the Triassic). Finding microvertebrates often requires hand quarrying, as with the dryolestid jaw I found in Wyoming, or screenwashing, a process that places soft-surface materials (or broken-down rock) in screen boxes in water to sieve out most of the bulk sediment and pick through "concentrate" under a microscope. Through these processes, we find many tiny fossils that simply couldn't be seen in the volumes of unconcentrated matrix.

These microvertebrates found by screenwashing or hand quarrying form the core of the nondinosaurian groups that we'll be focusing on here. I had been working with these fossils just a little for several years, but finding the gleaming mammal jaw in the Morrison Formation was the hook that drew me into the world of the wild beast companions for decades to come.

These microvertebrates and other contemporaries of dinosaurs lived long ago, in a world rather different from what we think is typical but is in fact just the current moment, geologically speaking. The world was in the process of becoming biologically modern during the Age of Dinosaurs, the Mesozoic era of Earth's history. Most animal groups familiar to us today began to appear then. As the Cambrian period 530 million years ago gave us the major animal body plans—the phyla—of today, the Mesozoic era (the Triassic, Jurassic, and Cretaceous) gave us many vertebrate groups and ecologies that we see in the world around us now. Turtles, lizards, crocodiles, mammals, birds, secondarily aquatic tetrapods, flying vertebrates—all of these appeared at some point during the 160-million-year reign of what everyone gets all worked up about—the dinosaurs (which admittedly were spectacular animals). And most of those specific groups actually outlived the dinosaurs. But dinosaurs aren't the whole story of the Mesozoic. In fact, in some ways, they were a relatively minor component of it, as all large vertebrates are, during their respective times, outweighed (literally) in terms of biomass and ecological influence by microbes, insects, and small vertebrates. And dinosaurs didn't operate as islands in their ecosystems. They were dependent on the other animals that lived alongside them. If the Upper Jurassic Morrison Formation of the western United States now has more than one hundred types of vertebrates known from its fossils, only about a third of those are dinosaurs—the rest belong to other groups we'll cover here.

But this was all quite a long time ago, an almost inconceivably long span by human timescales. It is the geological timescale and outline of the Mesozoic that we must tackle first. The staggering depth of geological time is something that might take a few false starts to eventually sink in with full impact. For me, that full impact hit one day while riding from San Diego to Utah.

Campanian Daydreams

I was slumped against the passenger side door, head on the inside of the window, staring down at the road rushing under my wheels you might say. I may have appeared asleep, but I was actually in a trance. At the wheel of the yellow International Harvester Scout was J. David Archibald, a biology professor at San Diego State University, barreling through the Mojave Desert northeast along Interstate 15 toward Nevada. Behind us in a white Ford Ranger were Matt Colbert (grandson of famous paleontologist Edwin Colbert) and Paul Majors, both paleontologists at the San Diego Museum of Natural History. Our destination was western Colorado, but I was lost somewhere in the Pleistocene.

Dave Archibald was a nearly-40-year-old paleontologist specializing in Mesozoic mammals and the Cretaceous-Tertiary extinction, as it was then known. A graduate of Kent State and the University of California, Berkeley (where he'd studied with Late Cretaceous mammal guru Bill Clemens), he had been studying the relatively unexplored paleontology of the Mesaverde Group of northwestern Colorado[1] for several years. He was headed out for five weeks of work with the three of us as field crew, all in our early to midtwenties. The Mesaverde there is a Late Cretaceous unit from about five million years before the end of the period, a stack of rocks several thousand feet thick in places that has older relatives of taxa like *T. rex*, *Triceratops*, and *Edmontosaurus* from the famous Hell Creek Formation in Montana and elsewhere. The importance of the Mesaverde in this area is that it sits geographically between deposits of similar age to the south in Texas, New Mexico, and southern Utah and those to the north in northern Montana and Alberta—and its fauna is not well known.

But I hadn't seen our field area yet. What I saw flying by on the pavement of the shoulder that morning was an imaginary geological timescale. Somehow, I'd seen in my mind a meter stick on the ground with a millennium marked off in 1000 millimeter-length years—a long human lifetime in about four inches. Then I imagined that geological timescale flying by below my passengers' side window at about 60 miles per hour. At that rate, we were traveling along my imaginary geological timescale back in time at about 26,400 years per second, about 95 million years per hour. All of the sunrises and sunsets, moon phases, seasons, eclipses, weather events, deposition, and erosion that take place in a single year were flying by at a rate of 1.58 million years each minute. It dawned on me that getting back to the beginning of the Cambrian period would require nearly another six hours of travel at the same rate. Leaving the Los Angeles basin around San Bernardino, our rate of travel back in time would put us in the later Early Cretaceous somewhere around Barstow in the Mojave Desert. By the time we passed through Baker, home of The Mad Greek and the World's Tallest Thermometer, still in the Mojave, we would be back in the time of the Chinle Formation, some of the world's earliest dinosaurs, and the giant logs of Petrified Forest National Park in Arizona. We wouldn't get back to the Early Cambrian until we were about two hours beyond Las Vegas, in the southwestern corner of Utah, near St. George. To get deep enough into the Precambrian to pick up the first unicellular organisms in Earth's history, we would have still needed to drive to a little beyond Chicago! The Precambrian comprises 88% of Earth's history, so most of the action we were after, although ridiculously ancient in reality, is geologically yesterday compared to the full geological timescale (fig. 1.1).

Although I'd recently graduated from college as a geology major, the mind-altering realization of the true length of geological time didn't really hit me until this trip. When I could see both a relatable length of time on the pavement (one year as a millimeter) and years fly by at such an incredible rate—and then understand how long it took to blaze through such a nearly incomprehensible number of years laid out end to end while Dave kept the pedal to the metal (on an International Scout nearly as old as I was)—that's when geological time began to work its psychedelic magic on my brain. One planet—but so many worlds.

The so-called Age of Dinosaurs, the Mesozoic era, ran from about 252 million years ago to about 66 million years ago (fig. 1.1), bookended by cataclysms. We'll get to that later, but this period of time was marked by the rise of dinosaurs (actually well into the Mesozoic), their dominance for more than 100 million years, and their rather dramatic demise. Or rather the demise of those unlucky enough to be around at the time—the vast majority of dinosaur species had in fact been extinct well before the disaster at the end of the Cretaceous. Most species last at most a handful of millions of years each, and dinosaurs had been around for just over 160 million years at that point, so a good 80–90% of dinosaur species that had ever existed were likely already extinct at the end of the Cretaceous period.

The Mesozoic started off with the Triassic period, which ran from 252 million years ago (mya) to 201 mya (fig. 1.1). These 51 million years included the appearance of dinosaurs about 230 mya and the dinosaurs biding their time among a host of large nondinosaurian reptiles that appear to have been more successful at being large reptiles than the dinosaurs were – for some 29 million years. By the end of the Triassic, the dinosaurs rose to dominate the landscapes. The Jurassic period (201–145 mya; fig. 1.1) saw this rise and the diversification and dominance of the dinosaurs to a peak of sheer size by the end of the Jurassic. The Cretaceous period (145–66 mya; fig. 1.1) saw some changes in what dominated the faunas in different parts of the

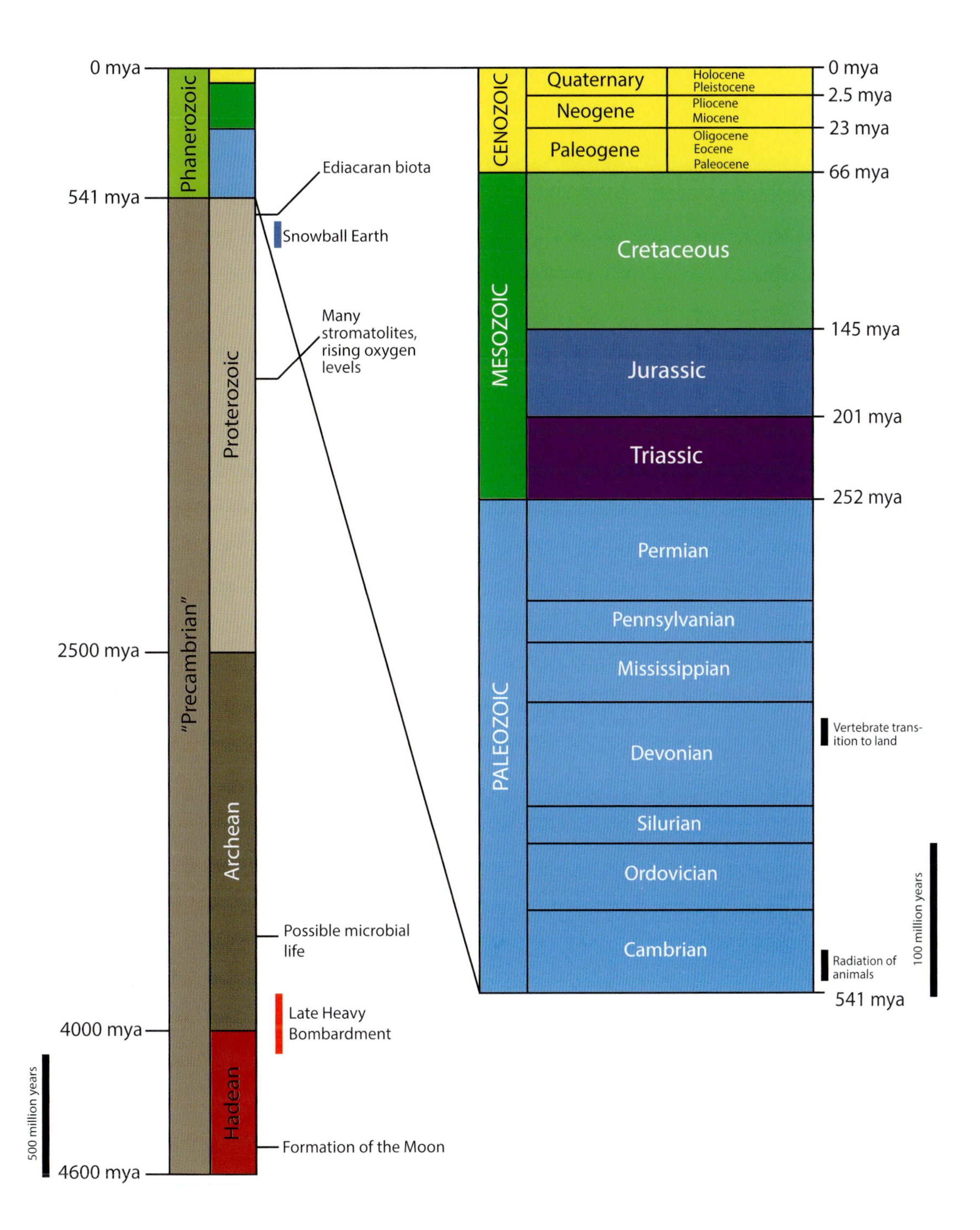

1.1 Geological timescale of Earth's history with subdivisions and to scale (except within the Cenozoic era enlargement where subdivisions are schematic). All of Earth's history on the *left*, Phanerozoic eon enlarged on *right* and showing subdivisions of the Mesozoic era, including the Triassic, Jurassic, and Cretaceous periods. Major events labeled. Note actual length of Precambrian compared to Phanerozoic on *left*. See figure 1.3 for evolution of groups within the Mesozoic (mya = millions of years ago).

world and included the time of the largest carnivores and some of the most sophisticated herbivores in dinosaur history, but it all ended rather suddenly and unexpectedly. Simple bad luck.

This book concentrates on animals that dominated the golden days of the dinosaurs during the Jurassic and Cretaceous periods. Many of the groups of animals common in those times have survived to today's world, and we will tackle some of the major groups in the following chapters. But first, we will explore a little of the weird, wild world of the Triassic—a time dominated not by dinosaurs but by unusual reptiles and amphibians that barely played much of a role (if any) throughout the rest of the Mesozoic. These groups of the Triassic, a time when dinosaurs still played a relatively minor role, will be the first stop of our journey.

Petrified Forests

The first period of the Mesozoic, the Triassic, produced the Chinle Formation of the southwestern US. This rock unit is the producer of all the giant logs of the Petrified Forest National Park in Arizona. The trees were araucarians similar to today's monkey puzzles and Norfolk Island pines, but the formation also contains leaf fossils of cycads. In fact, the Triassic landscapes of the planet were flush with mosses, primitive vascular plants, ferns, horsetails large and small, ginkgoes, cycads, conifers, and cycad-like bennettitaleans, most of which had appeared well before the start of the Mesozoic. By Jurassic times, this same cast of vegetation characters was still in place and in formations such as the Morrison Formation of the western US: diverse but open conifer forests sported trees up to 61 m (200 ft) tall and more. By the Late Jurassic, the very first angiosperms, or flowering plants, had probably appeared but not made much impact. The rise of these flowering plants throughout the Cretaceous period brought the first fruits, flowers, and grasses and facilitated a coevolution of plants and insects and plants and vertebrates that continues today, although with large mammals replacing dinosaurs.

Although it is not covered here, the story of the plants of the Mesozoic is an interesting and important one, and it is worth learning about in other sources. The dinosaurs and the beast companions of the Mesozoic food webs needed the energy of their systems to start somewhere, and these primary producers represented by the plant fossils are the foundation of all of the diversity at the higher levels. Without them, we'd have nothing to talk about here.

. . . Oh My

In southern Utah lies an isolated but beautiful stretch of Utah Highway 95 that runs between the towns of Blanding and Hanksville, passing the Bears Ears and Natural Bridges National Monuments and the northern stretches of Lake Powell. But along the way, the highway passes north of Red Canyon, hidden behind Wingate Mesa to the south. Once you have entered Red Canyon, however (if you happen to venture down gravel Uranium King Road or one of the other access points), you enter exposures of the

Late Triassic–age Chinle Formation that run for miles, almost literally for as far as you can see, topped by the Wingate Sandstone. The multicolored badlands and sandstone benches of the Chinle here are fertile ground for large and small animals that lived alongside the earliest dinosaurs found in North America. It is also fertile ground for uranium mining, with some of the few current (though dormant) mines in the country. But this was active real estate back in the 1950s when places like the Happy Jack Mine had whole communities spring up nearby for the families of the workers. It is a bit quieter now, but you see evidence of the old mining site and one or two current mines throughout the area. And the paleontological history of the area dates back a few years as well, as paleontologists like J. Michael Parrish and others collected material from this area back in the 1980s.

Wandering the rough terrain of this area in the heat of a July morning several years ago, I was off on my own, out of view of and separated by hundreds of meters from the rest of our crew, which consisted of Appalachian State University paleontologist Andy Heckert and his students. Andy is a friend and colleague known to hit a few baseballs and occasionally to turn a two-mile errand into an 11-mile motorcycle joyride for its own sake. He specializes in vertebrates of the Chinle Formation and operated out of New Mexico for years, usually sporting a foreign legion–style kepi in the field, and he now brings his students from Appalachian State to the Southwest to continue that exploration.

Walking along a flat, out from the base of the Petrified Forest Member (a stratigraphic subunit of a formation) of the Chinle, I was surveying the white surface of the flat scattered with hundreds of small pieces of petrified wood, trying to see if any fossil vertebrate material was mixed in with the wood, exposed and contrasting with the light background of the dirt. After several minutes, I was surprised to spot a tiny, comb-shaped something that stood out from everything else in both color and shape, although it was less than 5 mm long. When I laid down on the ground and carefully picked it up, I could see through a hand lens that it was in fact a piece of the tooth plate of a lungfish. Lungfish today are relatively robust fish that live in freshwater in tropical parts of Australia, South America, and Africa, but their history dates back to the Devonian period of the Paleozoic, a good 150 million years before dinosaurs arrived. They are members of the sarcopterygian lineage of bony fish, the somewhat less diverse branch opposite ray-finned fishes that we are more familiar with (fig. 1.2). Among the sarcopterygians are lungfish, coelacanths, and early prototetrapods—it is from this lineage that the first land vertebrates arose in the middle part of the Paleozoic era (figs. 1.1, 1.2, and 1.3). Lungfish are distinctive among fossils of the Mesozoic in that their jaws contained large tooth plates shaped like massive combs, with outward-facing ridges and an enamel surface pockmarked with pits. These are completely different from the long rows of sharp conical teeth that most ray-finned fish possess. What I now held between my fingertips as I was lying there on the southern Utah desert floor, seen by no one except perhaps a lizard, was most of a tooth plate of a Late Triassic lungfish, and one of the smallest I'd ever seen. A little air-breathing fish was among the

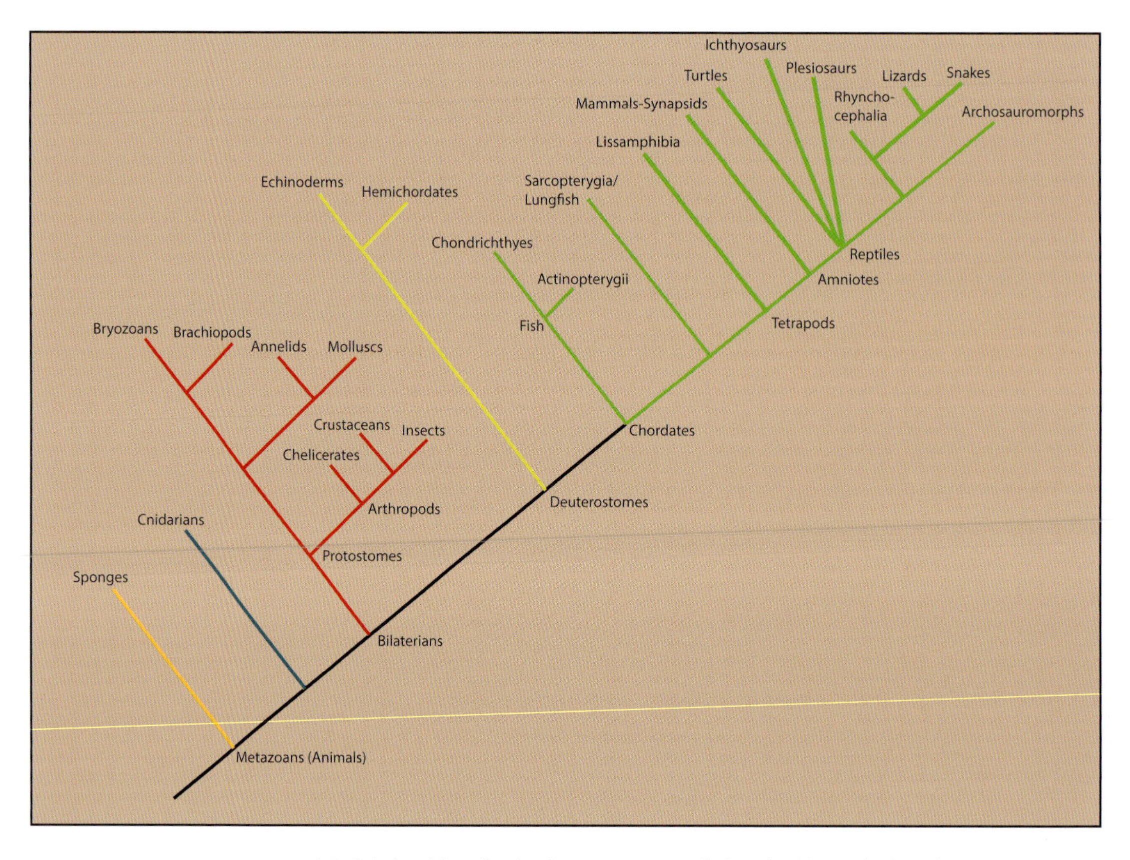

1.2 Relationships of animal groups present during the Mesozoic. Broader group names listed at node bases. Archosauromorphs include crocodiles, dinosaurs, and birds, among others. Chordate groups have green branches, other deuterostomes have yellow, protostomes have red, and cnidarians and sponges have blue and golden branches, respectively.

animals living in the lakes and rivers in what would become Utah in the same environment in which the first North American dinosaurs appeared.

But that little flesh-finned fish was nowhere near alone among the nondinosaurian vertebrates of the Late Triassic. Most of the vertebrate world of the Late Triassic consisted of nondinosaurs. In the waterways were lungfish, palaeoniscoids, semionotids, coelacanths, and sharks among the fish, along with giant carnivorous amphibians, metoposaurs, and even larger carnivorous phytosaur reptiles. On land, there were herbivorous aetosaurs and dicynodonts, carnivorous therapsids, rauisuchids, and early crocodilians. There were mammals and pterosaurs and lizards and a whole host of nonlizard but lizard-like reptiles. There were reptiles that looked a bit like dinosaurs but weren't. And then there were a few dinosaurs. All these

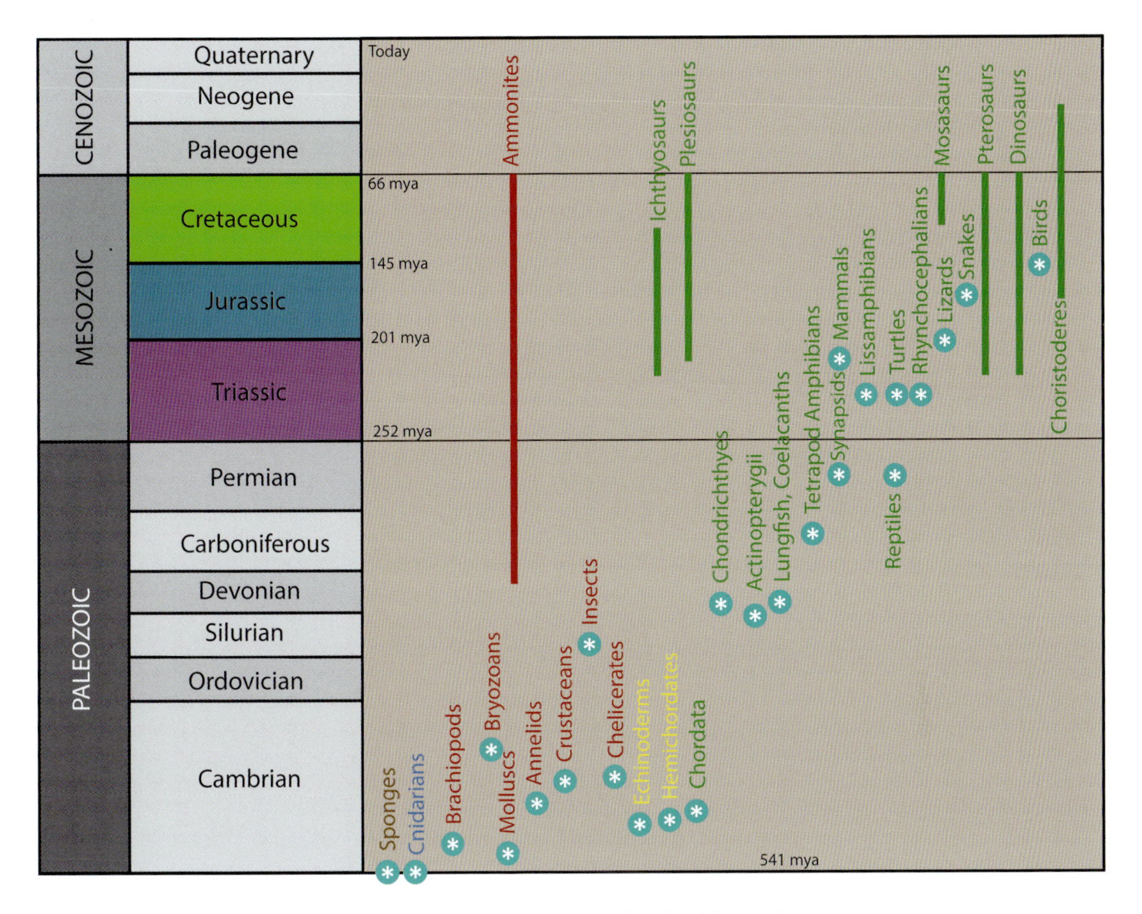

1.3 Ranges of animal groups on the Phanerozoic timescale. Asterisks (*) in blue circles indicate the points of origins of groups still represented in today's fauna (these groups have been present from the time of their appearance to modern times at the top of the chart). Colored bars indicate the ranges of extinct groups. Sponges in brown, cnidarians in blue, protostomes in red, invertebrate deuterostomes in yellow, and chordates and vertebrates in green. All major animal groups were present during all or some of the Mesozoic; also note that of later-appearing chordate groups, most appear near the same time as dinosaurs in the Triassic period.

occupied the world of the Late Triassic when Pangea was just beginning to split apart. The western United States' Chinle Formation was deposited in western Pangea at the time and straddled the northern tropics at about 5°–15° latitude. The Chinle landscape was well forested around lakes and rivers, but conditions became drier through the Late Triassic.

The Chinle Formation in Utah extends through the eastern part of the state from south to north, from the Four Corners up to Dinosaur National Monument. Around Moab, the formation consists of sandstones and mudstones sandwiched between the brick-red silts of the underlying Moenkopi Formation and the towering orange cliffs of the Wingate Sandstone. It was in this terrain in the Chinle west of Arches National Park that I set to work one afternoon years ago, prospecting for teeth and bones in the conglomerates of the formation exposed on the slope below the Wingate cliff. My volunteer Ray Bley and I set out up the slope looking in each block loose on the surface, in an area recommended to us by a friend and associate of our museum, the cast-dinosaur artist Rob Gaston. As I went up the slope, Ray sat down, pulled out hammer and chisel, and began banging away on one large chunk of conglomerate, chipping off pieces of rock in search of bones and teeth not otherwise exposed. I told Ray that there were plenty of blocks that could be checked on the surface (hundreds, really) without needing to pound. It is a good thing Ray ignored me, because not ten minutes later, he called me downslope from my search with a curious "What's this?"

Sure enough, Ray had found a bone inside his conglomerate block, and not just any bone but the fused pelvic vertebrae of a small coelophysoid dinosaur, the first found in Utah. Such dinosaurs are rare in the Chinle, not just in Utah but in the rest of the Four Corners too. I returned to that same canyon a number of years later and, on the same slope, found a small red sandstone slab with the three-toed hindfoot track and four-toed forefoot track of a small dinosauriform, a track type called *Atreipus*, which is common in the eastern United States but quite rare in the Chinle. The trackmaker probably belonged to a group called the silesaurids, which are nondinosaurian dinosauriforms known from the Late Triassic. Nondinosaurs that looked like dinosaurs might be a theme for the Late Triassic, as the reptiles that really make the epoch what it is are mostly on separate branches of the reptile tree from the famous "terrible lizards," as we will soon see.

Lungfish and small dinosaurs and silesaurids. This is just the tip of a surprising Triassic iceberg.

Triassic Park

The Triassic period of the geological timescale was named in 1834 for a sequence of beds in northern Germany. The base of the Triassic is now designated as the base of a particular bed on an outcrop in China, a level known as a Global Stratotype Section and Point (GSSP), a so-called golden spike that is in most cases neither golden nor a spike, more of a bronze cap. It is designated on the first appearance of a particular conodont species at a section near Lake Tai west of Shanghai.[2] Dinosaurs don't appear until the Late Triassic, about 230 million years ago.

The Triassic period began with complete disaster. An extinction so massive it nearly killed off all animal life in the oceans, and those animals on land fared only a bit better. Plants took a huge hit as well. The earliest Triassic was a time of acidic, anoxic ocean waters (except near the surface where oxygen was present), and of almost barren ocean floors. Marine mollusks, plankton, fish, and most other forms were far less diverse and far less densely distributed than they had been a few million years earlier. The culprit? For years, thanks to hip paradigms and the bandwagonism that even scientists are sometimes susceptible to, it was essentially assumed that it was an asteroid or comet even more massive than the one that ended the Cretaceous. But there are careers to be made by pointing out why and how the majority is wrong, and so when the majority is wrong in science, it generally gets corrected. For hundreds of years, nearly all geologists assumed that continents were stable and unmoving and that faunal and floral similarities on continents separated now by oceans in different times of Earth's history were due to almost magical land bridges that appeared between landmasses and then disappeared again. And then, after World War II, oceanographic work mapping the seafloors of the world's oceans demonstrated such overwhelming evidence for plate tectonics that in barely a decade all the previous assumptions were thrown out. With the evidence available, plate tectonics now seems almost intuitively obvious, but then so did stable continents in their day. Regardless, when old ideas turn out to lack evidence, the truth comes out sooner or later.

In the case of the Permian–Triassic extinction, a number of asteroid specialists, who had worked on the Cretaceous and other extinctions already, demonstrated that many of the telltale signs of impact simply weren't present at the Permian–Triassic boundary and that whatever caused the extinction, it was unlikely a comet or asteroid impact. Rather, the culprit appears more likely to have been volcanic eruptions in Siberia that pumped out so much carbon dioxide (CO_2) in such a short interval of time—and which coincidentally burned massive regional coal deposits in the process, releasing even more carbon—that the atmosphere warmed significantly and the oceans warmed and lost oxygen in their shallows. Compounding this, warming of the oceans then may have caused seeping of methane from deposits in the deep.[3] As a result, ocean temperatures near the equator may have hit levels you experience today in a hot tub—up to about 104°F (40°C)! These conditions appear to have been so different from previous ones and to have occurred so quickly, geologically speaking, that it proved too much for most marine and many terrestrial species and then ecosystems. The extinction at the Permian–Triassic boundary may have occurred in less than 60,000 years—about as close to geologically instantaneously as we can currently resolve.

Complicating the picture is that the type of volcanism might have made a difference in the case of the Permian–Triassic extinction. Around the time of the extinction, the Siberian eruptions appear to have switched from predominantly flood lavas flowing across the land surface to within-bedding, subsurface sills that more or less baked out methane and carbon dioxide from the native sedimentary rock into which the igneous magmas were intruding. It appears that the early lavas that cooled at the surface may have effectively capped and, through loading of their mass, pressured the dike and sill system below to spread out and more extensively cook the host rock, releasing additional greenhouse gases. As mentioned above, the extinction at this boundary appears to have occurred within a few tens of thousands of years, and although there may have been at least two pulses of extinction separated by about 180,000 years, there is no consensus on this.

Either way, ocean warmth also may have caused the acidification of marine waters that killed off a number of species of marine invertebrates, including corals, almost entirely. In fact, the only corals that survived did

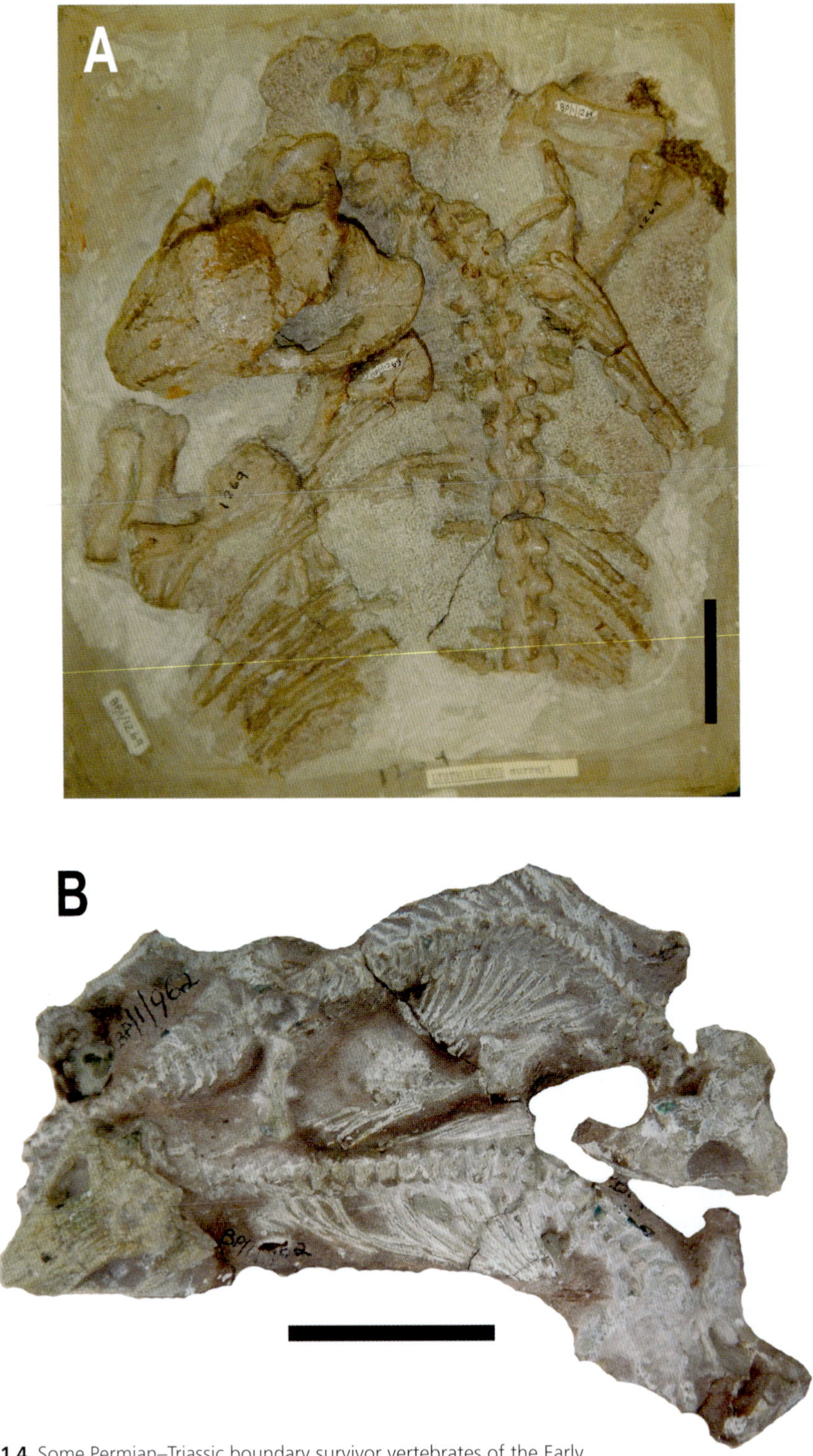

1.4 Some Permian–Triassic boundary survivor vertebrates of the Early Triassic from South Africa. (*A*) Skeleton of the dicynodont *Lystrosaurus* (BP/I 1269). (*B*) A pair of procolophonid (*Procolophon*) skeletons (BP/I 962). Both scale bars = 5 cm. *Photos courtesy of Andy Heckert.*

so living like sea anemones without their calcite cup skeletons, and they weren't able to begin making skeletons or to diversify into the scleractinian corals we know from today's reefs until several million years into the Triassic. Very warm temperatures, acidification, and lack of oxygen resulted in a worst-case combination for most marine animals.

The recovery from this extinction took a while to gather steam, but eventually, the Triassic world on land was well populated with survivors such as little *Lystrosaurus* and the procolophonids (fig. 1.4) as well as new forms that were beginning to appear. What proved to be the age of large reptiles ruling the land and seas during the Mesozoic was set to take off, even if this would not have been obvious to any of the animals present in the middle years of the Triassic.

Triassic Large Archosaurs

Once the Triassic got going, however, the "ruling reptiles" eventually took off. By that, I mean the archosaurs, not the dinosaurs specifically. The Triassic was a truly atypical time in the Mesozoic when the dinosaurs took a back seat in terms of diversity and numbers—and even size and flashiness—to the nondinosaurian archosaurs. Rauisuchids, phytosaurs, and even aetosaurs were all large, often armored reptiles that were larger, and in some ways more interesting, than their dinosaurian counterparts.

Reptiles appeared early in tetrapod history, but their full modern diversification took place during the early to middle part of the Triassic period. As we will see later, lizards, sphenodontians, and turtles originated around this time too, but the archosaurs were the real story of the time. Not long after their appearance, archosaurs split into the Pseudosuchia and the Ornithodira (fig. 1.5). Phytosaurs may be just outside Archosauria. The pseudosuchians include the aetosaurs, rauisuchids, and crocs, while the ornithodirans consist of archosaurs closer to birds than to crocs and include the pterosaurs, silesaurids, nonavian dinosaurs, and avian dinosaurs (i.e., birds). The phytosaurs, rauisuchids, and aetosaurs are the stars of the Triassic. If you've ever vacationed in northern Arizona and on your way to Grand Canyon stopped by Petrified Forest National Park, you have seen the rocks preserving the Late Triassic ecosystems of the western United States in the Chinle Formation. Fossils of phytosaurs and aetosaurs are common in this terrain, which is mostly famous for the giant petrified trees dominating parts of the landscape. Rauisuchids are much less common, and dinosaurs are rarer still. This is the key pattern of the Late Triassic when the dinosaurs were a minor component and the world belonged to other archosaurs. Phytosaurs were giant, semiaquatic, and ecologically crocodile-like predators that prowled the lakes and rivers of the time (figs. 1.6 and 1.7*A*). Rauisuchids were just-as-wicked terrestrial predators with skulls and teeth superficially similar to those of the later, large theropod dinosaurs (fig. 1.7*D*). Aetosaurs, meanwhile, were heavily armored plant eaters or omnivores with bony plates in their skin that beat the dinosaurian ankylosaurs to the niche punch by some 70 million years (fig. 1.7*B* and 1.7*C*). Cast skeletons of some of these animals are displayed at the Rainbow Forest Museum at Petrified Forest.

1.5 Relationships of the "ruling reptile" archosaurs showing the Ornithodira in yellow and the common Triassic Pseudosuchia, which were more dominant than dinosaurs before the Jurassic, in blue.

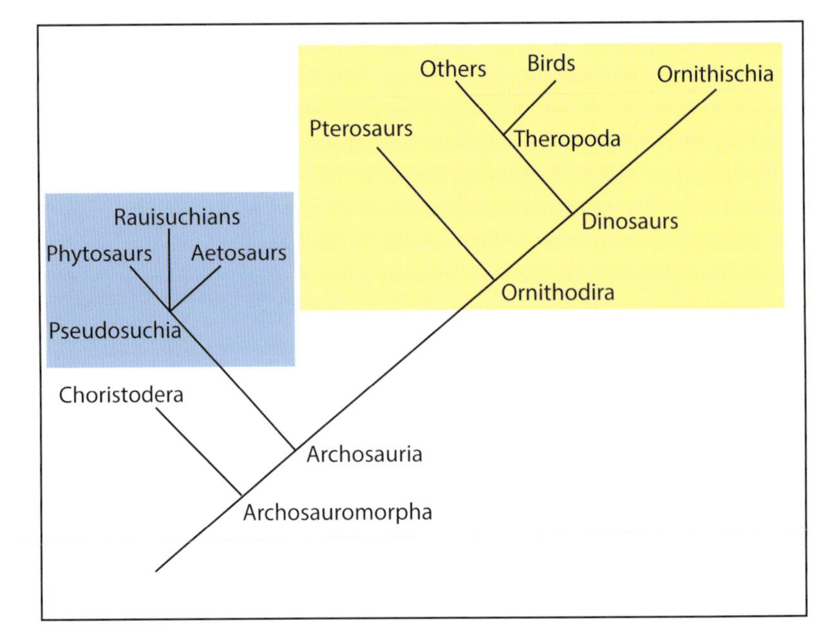

1.6 Phytosaur fossils of the Late Triassic–age Chinle Formation of the southwestern United States. (*A*, *B*) Skulls of the phytosaur *Pseudopalatus* from New Mexico. (*A*) Male (*top*, NMMNH P-31292) and female (*bottom*, NMMNH P-50040) from the Petrified Forest Member. (*B*) NMMNH P-4983 (*top*) and NMMNH P-4256 from the Bull Canyon Member. (*C*) Phytosaur trackway (*Apatopus*) from the uppermost Chinle just outside Capitol Reef National Park, Utah. Note two left manus-pes sets just left of tail drag and 10-cm scale bar. Tracks include scale skin impressions and claw drag marks. Surface also features numerous small reptile tracks assigned to *Gwyneddichnium* and *Rhynchosauroides*. This track slab is now at the Museums of Western Colorado. *All photos by author; A and B courtesy of New Mexico Museum of Natural History and Science.*

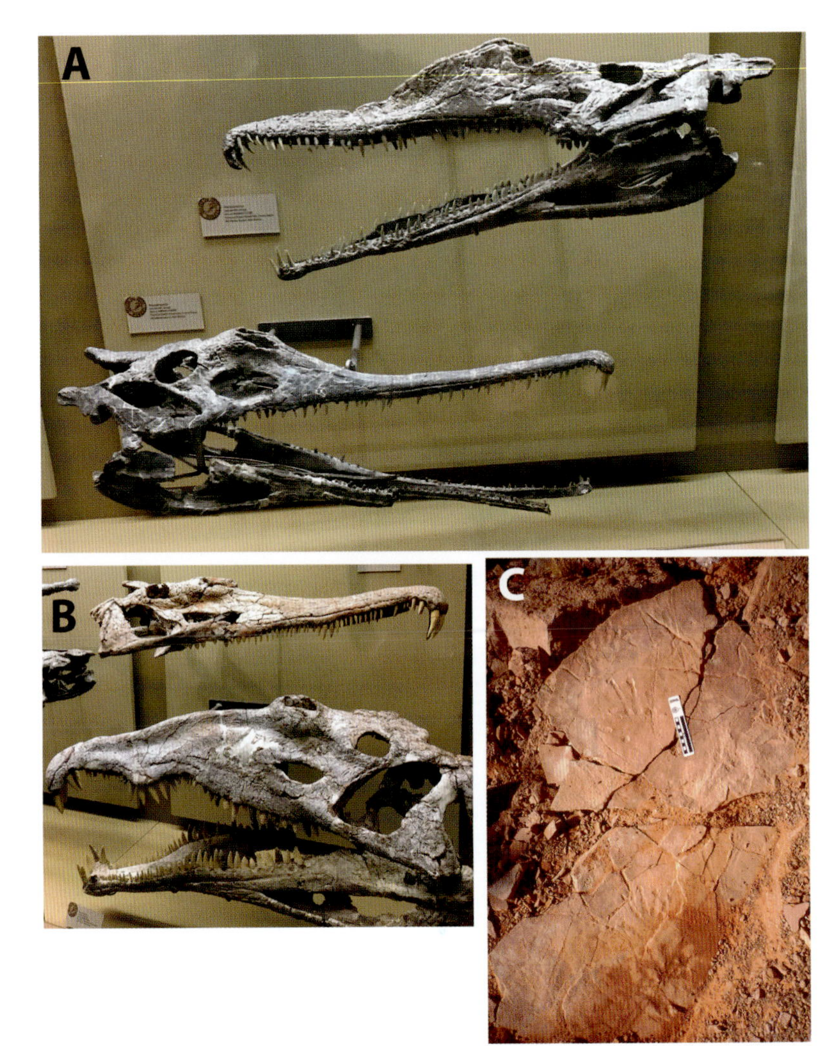

1.7 Large archosaurs and a synapsid from the Chinle Formation of the southwestern United States. (*A*) The phytosaur *Smilosuchus*. (*B*) The aetosaur *Typothorax*. (*C*) The aetosaur *Calyptosuchus*. (*D*) The rauisuchian *Postosuchus*. (*E*) The dicynodont synapsid *Placerias*. *All images courtesy and © of Jeff Martz.*

Other museums also have mounts of these Late Triassic animals, and many of the casts are based on material from Petrified Forest. Wherever you find them, they're worth a good long look because to some degree they anticipate later reptile designs—the phytosaurs beating the later crocodiles to the semiaquatic ambush predator party; the aetosaurs, as mentioned, being a step or two ahead of the ankylosaurs in the armored plant-eater department; and the rauisuchids being (very broadly) facultatively bipedal equivalents of the later, larger theropods.

Another group of large nondinosaurians abundant at various times and places were the dicynodonts. In the Chinle Formation of western North America, dicynodonts occur in a few places but are perhaps most abundant in a quarry southeast of Petrified Forest National Park where the form *Placerias* has been found (fig. 1.7*E*). Tracks of dicynodonts are also known from Shay Canyon in the Chinle on the way to the Needles District of Canyonlands National Park.

Dicynodonts were very abundant in the Permian; at least, their fossils are commonly found there. But the group took a big wallop in the Permo–Triassic extinction. Almost three dozen genera of dicynodonts lived during the Late Permian, but only two survived into the Early Triassic. Among the survivors was little *Lystrosaurus*, who has been found in the Permian to Early Triassic of South Africa, Antarctica, India, and Asia. Given its distribution and the current location of the continents on which it occurs, *Lystrosaurus* played a role in early suspicions of "continental drift," which was eventually shown to be likely true by way of plate tectonics. *Placerias*, from the Late Triassic, was one of the largest (and last) genera of dicynodonts on the planet; most Permian forms were smaller and relatively more slender.

In the Paleozoic era, the first land vertebrates were amphibians, but a group of these developed an enclosing sac for the embryo within their eggs that allowed them to lay their eggs away from water. This group was the amniotes and includes modern reptiles, birds, and mammals. Early in amniote history, a group known as the synapsids split off and eventually led to mammals. Dicynodonts of the Late Triassic are a synapsid holdover from the Paleozoic. *Placerias* was big, with a boxy head, a large beak, two large tusks, a short, heavy body with relatively stout limbs, and a short tail. It is best known from the Chinle Formation of Arizona and was much heavier than most dinosaurs of the time. It probably rooted around for plants for food. Dicynodonts disappeared at the Triassic–Jurassic boundary.

These four groups of large animals (dicynodonts, aetosaurs, phytosaurs, and rauisuchids) lived alongside actual dinosaurs, but the dinosaurs of the Chinle Formation were a rather small, slender, and rare lot, typified by the famous *Coelophysis*, best known from nearly 200 skeletons at Ghost Ranch, New Mexico. This carnivore was dwarfed by the phytosaurs, dicynodonts, and rauisuchids and would have been no danger to an adult aetosaur, so between numbers and diversity, the Chinle dinosaurs were not a huge factor in the ecosystems.

Fossils of phytosaurs, aetosaurs, dicynodonts, rauisuchids, and rare dinosaurs range from this Petrified Forest part of northeastern Arizona east into New Mexico and north and west into Colorado, Utah, and Wyoming, but some are also known from the eastern United States and elsewhere in the world.

Triassic Small Oddballs

Microvertebrates are found, sometimes in great abundance, in the Upper Triassic formations of the southwestern United States, and they are often found by screenwashing or hand quarrying. Many of these specimens are isolated small bones or teeth that don't appear to belong to any obvious group, and although many have been assigned taxonomic names, we're not always entirely sure from what animals the elements derived. Rock units such as the Chinle and Dockum and some of their subunits such as the Petrified Forest, Colorado City, Tecovas, San Pedro Arroyo, and Blue Water Creek have yielded many teeth and small bones representing a wide range of animals, including sharks, fishes, metoposaur amphibians, procolophonids, synapsids like *Adelobasileus*, rhynchocephalians, *Trilophosaurus*, phytosaurs, aetosaurs, and dinosaurs, among the latter theropods, prosauropods, and possible ornithischians. And such groups can be found in Late Triassic–age rocks in many parts of the world, along with unique forms, as we will see.

Procolophonids were small reptiles with stout bodies, spiked skulls, and short tails; *Hypsognathus* from the Late Triassic of New Jersey, Connecticut, and Nova Scotia was especially spiked along the cheeks on the side of the skull and was about 12 inches long. Most Late Triassic procolophonid forms were herbivorous, but the group started out in the Early Triassic, evolved from insectivorous Carboniferous to Permian ancestors. The group was once thought to be related to turtles, but now (with turtles being modified diapsids) they are considered "parareptiles," or primitive reptilians outside both turtle and diapsid groups.

The Triassic had a number of other rather offbeat nondinosaurian reptiles inhabiting its forests, floodplains, and deserts. At the time of the dinosaurs' first appearance, there seem to have been so many reptiles of so many new designs that few people would have selected the dinosaurs as the group that would eventually take over the terrestrial world and rule it for more than 130 million years. But the weird reptiles of the Triassic were numerous and widespread. Most were relatively small; it was a good idea to stay off the radars of the rauisuchids and phytosaurs.

Trilophosaurus was a komodo dragon–sized, lizard-like, herbivorous archosauromorph (a group including archosaurs and their more primitive immediate relatives) from the Chinle Formation with teeth possessing several mediolaterally oriented ridges. This would be an odd combination for modern reptiles—big and herbivorous—but it was, after all, the Late Triassic and normal rules of later years did not yet apply. The sister group to trilophosaurids, the azendohsaurids *Azendohsaurus* and its relatives, consists of Middle to Late Triassic herbivorous taxa from Morocco, Madagascar, and India. These latter forms had leaf-shaped teeth rather different from the wide-ridged teeth of trilophosaurids. The Middle Triassic azendohsaurid *Shringasaurus* from India boasted horns over the orbits and had robust forelimbs. It was also relatively large as herbivores of the time went.

Drepanosaurus from the Late Triassic of Italy and New Mexico (fig. 1.8*A*) lived arboreally, with grasping hands and feet for gripping branches and possibly a tail-tip claw (fig. 1.8*B*), suggesting possible prehensility and the nickname "monkey lizards" for these animals. Although several other genera of drepanosaurs, such as *Megalancosaurus* (fig. 1.8C), also appear to have been arboreal, others, including *Hypuronector* from the Late Triassic

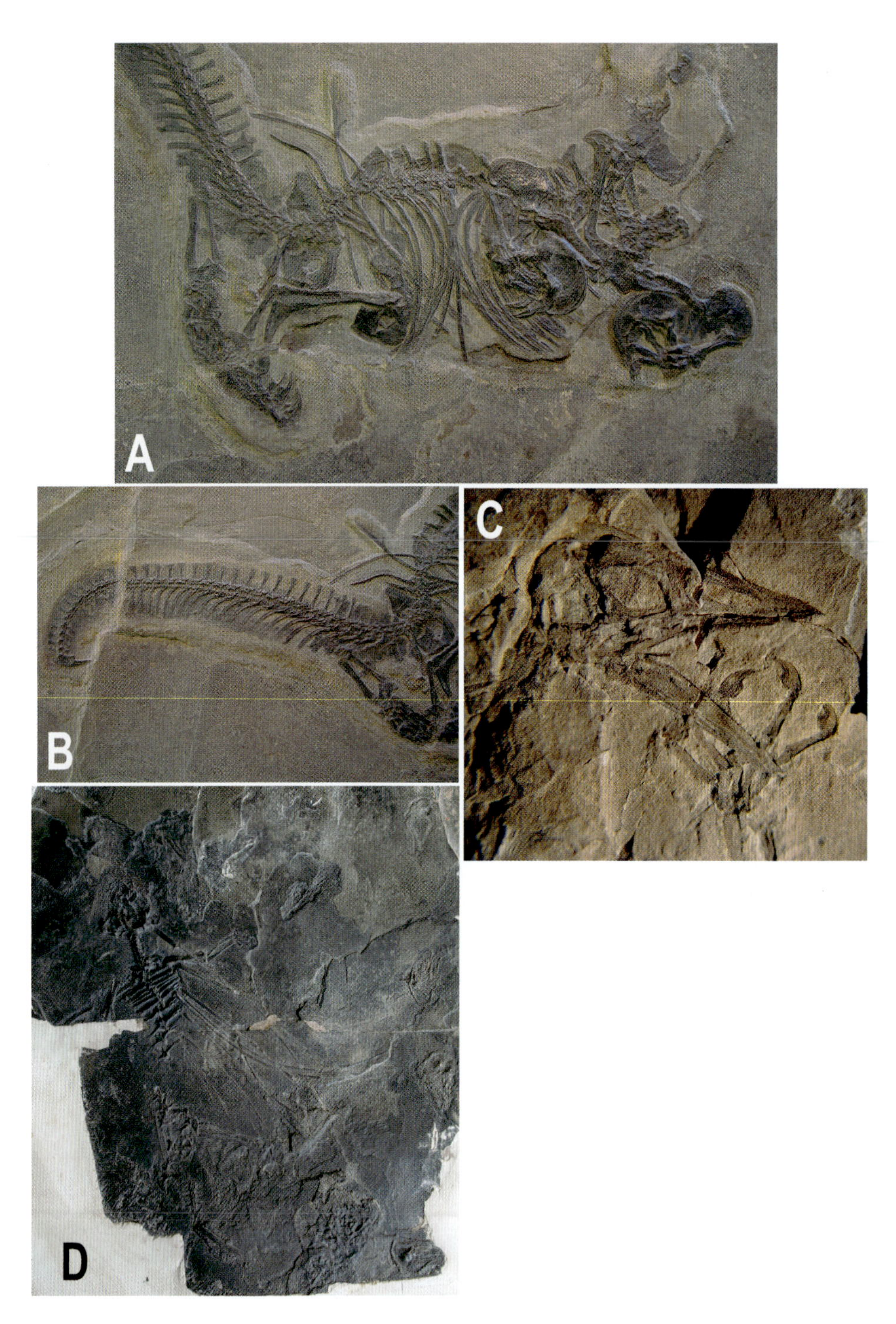

1.8 Unusual Triassic reptiles I. (*A*, *B*) The drepanosaur *Drepanosaurus* from the Late Triassic of Italy, showing the body and limbs (*A*) and the tail with an apparent distal hook (*B*). (*C*) Skull and manus of the drepanosaur *Megalancosaurus*, also from the Late Triassic of Italy. (*D*) The kuehneosaurid *Icarosaurus* with elongate rib extensions indicating a possible gliding capability (AMNH 2101). This reptile is from the Late Triassic–age Lockatong Formation of New Jersey. *All images courtesy of Nick Fraser.*

of New Jersey, had many deep caudal vertebrae, creating a ribbonlike tail, which suggested that this form may have been semiaquatic and that drepanosaurs may have been more ecologically diverse than we realized.

Kuehneosaurus, from the Late Triassic of Europe, was a gliding reptile with very elongate ribs and a skin membrane forming the stationary "wings" on either side of the torso. This reptile may have been capable of leaping from tree to tree or possibly from trees to the ground. It was not alone, either. The kuehneosaurid *Icarosaurus* from the Late Triassic–age Lockatong Formation of New Jersey also possessed elongate "gliding" ribs (fig. 1.8*D*). The archosauromorph *Mecistotrachelos* also glided thanks to elongate ribs (of different morphology than kuehneosaurids) and was found in the Upper Triassic Cow Branch Formation along the Virginia–North Carolina border.

Longisquama was found at what is largely an insect locality in the Middle–Late Triassic of Kyrgyzstan (fig. 1.9*A*). It was a small reptile with long feather-shaped scales along the back (apparently); these modified scales were much longer than the body. The question is whether these scales were arranged in a single line or in pairs. And were the scales lined up on either side of the trunk (as originally proposed) or along the back midline (as thought now)?

Tanystropheus was from the Middle to Late Triassic of Europe and Asia. It had an almost absurdly long neck with 12–13 extremely elongate vertebrae that together were several times longer than the trunk of the body (fig. 1.9*B*). This (along with the Paleozoic caseids[4]) was one of the first real weirdos I encountered as a young undergraduate geology student when it was featured in Robert Carroll's *Vertebrate Paleontology and Evolution*. This was one of those paleontological oddities that I had never seen before, despite a childhood full of dinosaur books. It made me think: *What on Earth is* that *thing?* You'd think a neck that much longer than the body would be a liability in the Mesozoic world, but apparently not. *Tanystropheus* was probably a semiaquatic piscivorous predator (fig. 1.9C), and it grew up to 20 feet long (though most individuals were smaller), with roughly half of that length being neck. The tanystropheid relatives *Tanytrachelos* and *Langobardisaurus* are known from the Chinle Formation in New Mexico and Arizona (and originally Virginia and Europe, respectively).

Tanystropheids may have made the small fossil footprints known as *Gwynnedichnium*, which are found in Late Triassic rocks in Europe and North America. I spent one summer solstice years ago in the Circle Cliffs of southern Utah, just west of Capitol Reef National Park, collecting a phytosaur trackway from the top of the Chinle Formation just below 122 m (400 ft) of Wingate Sandstone cliffs. This trackway had been found by Alden Hamblin and consisted of two left fore-hind footprint pairs and one right set. A tail drag mark wound through the middle of the trackway and the bottoms of the front and back feet left fine-scale skin impressions in the mudstone that preserved the trackway. Most interestingly, the fourth digit of the left hind foot appeared to be injured or malformed, missing the distal two phalanges of the toe. As we excavated more of the trackway and prepared to remove it, we discovered that the mudstone surface was also covered with

1.9 Unusual Triassic reptiles II. (*A*) The Middle–Late Triassic *Longisquama* from the Madygen Formation of Kyrgyzstan (PIN RAS 2584/4). (*B*) Reconstruction of the Middle–Late Triassic *Tanystrophaeus* from Europe (as displayed at the Paleontological Institute and Museum of the University of Zurich), showing the extremely elongate neck of this animal. (*C*) Skull of *Tanystropheus* (PIMUZ T2819) showing numerous sharp, conical teeth. *All images by Nick Fraser.*

1.10 Unusual Triassic reptiles III. (*A*) The semiaquatic *Vancleavea*, a form that may have reached lengths over 3 m (10 ft). (*B*) *Revueltosaurus*, a herbivorous pseudosuchian approximately 1 m (3.3 ft) long. *Both images courtesy and © of Jeff Martz.*

more than 50 tiny reptile tracks. Many of these were of a lizard-like animal with a track type known as *Rhynchosauroides*, but a number of them were the different form, *Gwynnedichnium*. These possible tanystropheid tracks are known from the Chinle in Utah, as we saw on this trip of long days, but they are also particularly common in the formation in Colorado.

Uatchitodon is an apparent archosauromorph known only from its teeth from the Late Triassic–age rocks of Virginia, North Carolina, and Arizona. The teeth are elongate, serrated, and slightly recurved, but more importantly, they have grooves along both the outer and inner surfaces of their lengths, and these grooves are either open or mostly enclosed depending on the species. These grooves suggest that the animal was venomous, as the grooves are generally similar to those of venomous snakes and the Gila monster lizard. This taxon was among the earliest venomous vertebrates, although earlier forms from the protomammalian line seem to have developed this ability by late in the Paleozoic.

Vancleavea is another form from the Late Triassic Chinle Formation of Arizona and New Mexico. It was armored, with short limbs, a long body, and a dorsoventrally deep tail (due to elongate osteoderms rather than neural spines) (fig. 1.10A). The flattened tail was probably used for propulsion, so the animal was likely semiaquatic. Years ago, I saw one of the better specimens of this taxon just exposed and still in a field jacket from the *Coelophysis* quarry when I visited with Alex Downs at the museum at Ghost Ranch, New Mexico. The armor-covered body looked like a medieval knight covered in chain mail and was impressively strange as Alex showed it off during a break in his prep work.

Revueltosaurus was a small, 1-m (3 ft) pseudosuchian reptile (fig. 1.10*B*) with teeth shaped like later herbivorous dinosaurs, and it likely ate plants as well. In fact, when first found (from its teeth), it was thought to be an ornithischian dinosaur. It is now known from several localities in the Chinle Formation and from North Carolina.

Pistosaurus and *Augustasaurus* from the Middle Triassic of Germany and France and then Nevada, respectively, were relatively small aquatic reptiles of the Pistosauria, probable primitive relatives of the later marine reptiles known as plesiosaurs. They had long necks, small skulls, and short tails, and their limbs were modified into paddles, but the bones of those limbs were not as specialized for their paddle-like structure as those of the later plesiosaurs would eventually be. We will get to the descendants of the pistosaurs in a later chapter.

The Middle Triassic of China produced another aquatic (probably marine) reptile known as *Atopodentatus*. These reptiles have a greatly laterally expanded front of the mouth, with an extremely broad row of pencil-like teeth at the front of the jaws (fig. 1.11A). The rest of the skull is fairly standard and almost rectangular in top view, but the mouth flairs out to the sides as if the animal were holding a giant harmonica just inside its mouth. What was this reptile doing with such a broad, straight row of small teeth at the front of its mouth? Grazing algae, apparently, making it one of the world's first-known herbivorous marine reptiles.

Also known from the Middle–Late Triassic of Kyrgyzstan, the small reptile *Sharovipteryx* had a somewhat unconventional morphology as well. The neck was slightly longer than the trunk, and the skull was relatively small, elongate, and triangular, the tail very long and thin. But the hind limbs were much longer than the forelimbs, and both appear to have had a gliding membrane (fig. 1.11*B*). In the forelimbs, this membrane probably stretched from the wrists back to just in front of the pelvis; on the hind limbs, where the membrane is preserved more clearly, the membrane seems to stretch all along the trailing edge of the upper and lower legs and from the inner toe to the base of the tail. The membrane may also have stretched forward from the knees to the midtrunk region, in which case, the forelimb membrane probably did not extend very far posteriorly. Because its forelimbs were so much shorter than the hind, *Sharovipteryx* appears to have been capable of gliding largely on its hind-limb membrane, possibly steering with the

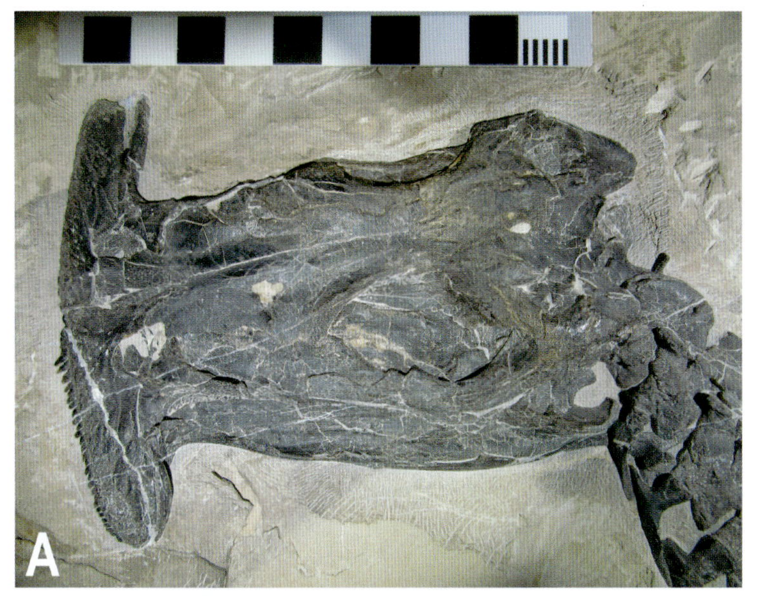

1.11 Unusual Triassic reptiles IV. (*A*) *Atopodentatus* from the Middle Triassic of China, demonstrating the broadly expanded mouth with tiny peg teeth of this marine reptile. (*B*) The Middle–Late Triassic *Sharovipteryx* from the Madygen Formation of Kyrgyzstan (PIN RAS 2584/8). Note the gliding membrane between the elongate legs and tail. Scale in *A* in centimeters. *Both images courtesy of Nick Fraser.*

head, neck, and forelimbs. The large-wing-in-back arrangement made the animal somewhat reminiscent of a tiny reptilian Concorde of the Triassic (but slower). Unfortunately, this taxon is represented by only a single specimen from the lake and river deposits at the type locality in Kyrgyzstan, and we know of only one similar reptile during the Late Triassic, the recently described *Ozimek* from Poland, so our understanding of these apparent gliders remains limited. It appears that *Sharovipteryx* and *Ozimek* may be archosauromorphs and are possibly related to tanystropheids.

The Late Triassic archosauromorph *Triopticus*, from west-central Texas, was dome-headed with thick bone forming the top of the skull. This morphology reflected but predated that of the unrelated pachycephalosaurid dinosaurs of some 100 million years later. In fact, as aetosaurs broadly anticipated the morphology of later, but unrelated dinosaurian ankylosaurs, and as phytosaurs preceded crocodilians as long-snouted, semiaquatic, carnivorous quadrupeds, *Triopticus* foreshadowed the pachycephalosaurs in the dome-headed niche. The gracile, bipedal to facultatively quadrupedal silesaurid dinosauriforms of the Triassic filled a similar role regarding the early ornithischian dinosaurs of the Early to Middle Jurassic. The beaked, bipedal shuvosaurids were to some degree the ecological predecessors of the unrelated dinosaurian ornithomimids of the Cretaceous. All of these convergent pairs demonstrate the pattern of dinosaurs and crocodiles of the Jurassic and Cretaceous, filling ecological roles left empty by their nondinosaurian distant relatives of the Triassic, after the Triassic–Jurassic boundary.

Triassic Summary

The Triassic was a time of transition, not just in the land animals but in the marine fauna as well. The Permo-Triassic extinction had dealt a big hit to the groups common in the Paleozoic, and the Triassic was a time when the modern marine invertebrate taxa rose to dominate the Mesozoic through today. On land, the reptiles of the Triassic flourished between the extinctions bookending either side of the period, with many of the odd groups and taxa we have just discussed appearing and creating the stand-out time in the Mesozoic—the Middle to Late Triassic—when many modern terrestrial vertebrate groups appeared at the same time as many unusual ones that were often extinct by the end of the period. The survivors of the Triassic–Jurassic boundary diversified to inaugurate the Mesozoic of the Jurassic and Cretaceous, a time dominated by the dinosaurs (just one of the groups that happened to survive the Triassic–Jurassic boundary) as well as many of the broad nondinosaurian vertebrate groups common today. So the Triassic was, from our perspective among today's fauna and from the hypothetical perspective of a late Paleozoic vertebrate looking forward, a long, strange trip of weird reptilian overlords (phytosaurs, rauisuchids, aetosaurs), bizarro small reptiles that glided on ribs or hind legs or had necks like a cartoon character, a few Paleozoic holdovers (dicynodonts, metoposaurs), and barely noteworthy groups like dinosaurs and mammals. This was a very different terrestrial world from the Permian and previous periods and one that no one could have predicted would soon be a distant memory after the beginning of the Jurassic when dinosaurs would take over essentially all of the large terrestrial vertebrate ecological roles and the nondinosaurian smaller vertebrates would consist mostly of modern groups such as lissamphibians, turtles, lizards, crocodilians, and mammals. All of these latter groups appear on the Triassic scene around the same time as the dinosaurs. All of these future stars of the Jurassic and Cretaceous, however, were born into the world of the nondinosaurian archosaurs and other strange beasts and had to hang back in the shadows, so to speak (or perhaps literally), for 30 million years or so before they took over the world themselves.

Jurassic Lark

Now, as we head into the groups that dominated the nondinosaurian world of the Jurassic and Cretaceous in the coming chapters, we do not actually leave the Triassic behind. As we will see, many of these groups had their origins in or even before the Triassic. We have merely set the stage. But before we venture through the rest of the Mesozoic and its cast of characters, here is a bit of background on the rest of the era, the Jurassic and Cretaceous periods.

More extinction hit at the Triassic–Jurassic boundary with an ecosystem collapse somewhat similar to the Permian–Triassic boundary, but it was smaller in scale and was probably caused by rapid warming and ocean anoxia brought on by volcanic activity in the Central Atlantic Magmatic Province. This volcanic activity was related to the start of rifting through the center of Pangea, a process that began the separation of Africa and Europe from South and North America. By the opening years of the Jurassic, the aetosaurs, phytosaurs, and rauisuchids were gone, and the dinosaurs, pterosaurs, and crocodylomorphs had survived, dinosaurs diversifying through to the Middle Jurassic to take on the now-unoccupied niches. Soon dinosaurs were in many cases far larger and as a group more ecologically varied than during the Triassic. Still, there were plenty of small dinosaurs mixed in with the tiny nondinosaurian taxa, and these turn up frequently in our hunts for microvertebrates, usually as teeth and fragments in screenwashed samples among many other types of vertebrate fossils (fig. 1.12).

The Jurassic was named for the Jura Mountains along the French–Swiss border. Deposits of this age occur on all continents on Earth, including Antarctica, home to the theropod dinosaur *Cryolophosaurus*. By the middle of the Jurassic, some theropods had gotten quite large, much larger than little *Coelophysis* from the Late Triassic. More advanced behemoth sauropods, armored-tank ankylosaurs, and spiked-and-plated stegosaurs had evolved among dinosaurs by this epoch, which more or less ushered in the beginning of the Age of Dinosaurs most people are familiar with. The Jurassic period lasted from 201 to 145 million years ago, and it brought the pinnacle of all weird dinosaur ecosystems in the time of the Late Jurassic—the epoch of so many of the absurdly large herbivorous giants in the diplodocoid and macronarian and turiasaurian sauropods, of single environments with multiple species of multiton bipedal carnivores, and of armies of low-browsing plant eaters of all varieties, bipedal, armored, and spiked.

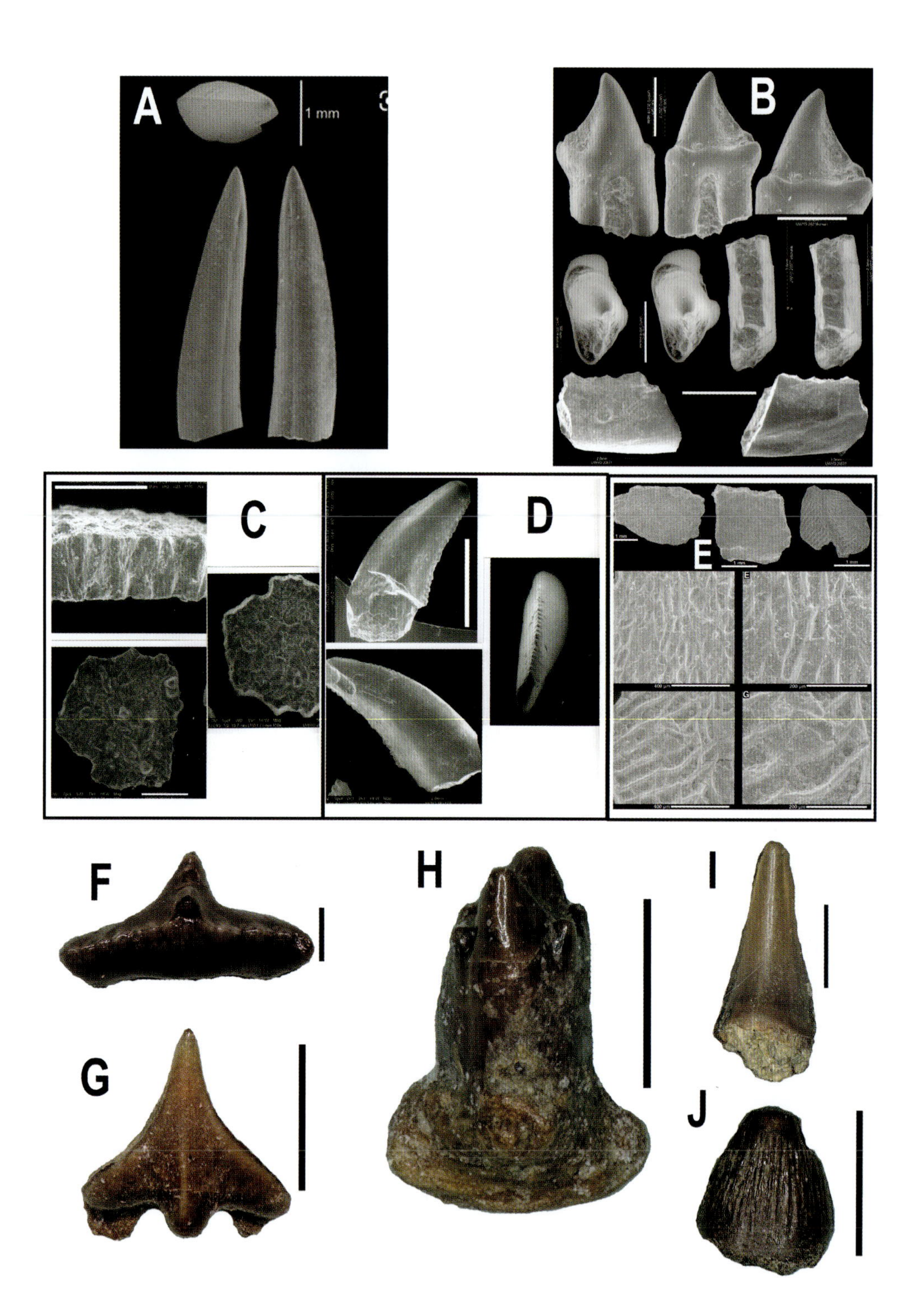
A
1 mm
B
C
D
E
F
G
H
I
J

Facing, **1.12** Images of vertebrate microfossils obtained through screenwashing of matrix from various Mesozoic formations in North America. (*A*) Tooth of the Triassic reptile *Uatchitodon*. (*B*) Multiple-view scanning electron microscope (SEM) images of a mammal premolar and jaw fragment (without teeth) from the Upper Jurassic Morrison Formation of northeastern Wyoming. Scale bars = 500 μm for premolar and 2 mm for jaw fragment. (*C*) SEM views of a dinosaur eggshell fragment from the Morrison Formation. Scale bars = 500 μm. (*D*) SEM views of a small theropod dinosaur tooth from the Morrison Formation. Scale bars = 2 mm. (*E*) SEM views of fragments of fish scales from the Morrison Formation. Scale bars as marked. (*F*) Tooth of the chondrichthyan *Lonchidion* from the Upper Cretaceous Williams Fork Formation of northwestern Colorado. Scale bar = 500 μm. (*G*) Tooth of the chondrichthyan *Chiloscyllium* from the Williams Fork Formation. Scale bar = 1 mm. (*H*) Tooth of the teiid lizard *Peneteius* from the Williams Fork Formation. Scale bar = 1 mm. (*I*) Fish tooth from the Williams Fork Formation. (*J*) Tooth from a small crocodilian from the Williams Fork Formation. Scale bar = 5 mm. *All images by Andy Heckert or Nick Brand and from Foster and Heckert (2011) or Brand et al. (2022).*

Camp Cretaceous

The Cretaceous was nearly as long as the Triassic and Jurassic combined, lasting from 145 mya to 66 mya, and was named after chalk deposits in northern France. Early in the period, a transition occurred from ecosystems dominated by sauropods, ankylosaurs, and allosauroids, among dinosaurs, to those featuring hadrosaurs and tyrannosaurs and eventually ceratopsians. That was in North America at least; elsewhere in the world, the sauropods hung in there in the form of massive titanosaurs. Asia and North America became the land of the tyrannosaurs, as the southern continents were more dominated by abelisauroid theropods. Late in the period, the hadrosaurs and ceratopsians became particularly diverse, and there was added variety in the pachycephalosaurs, ankylosaurs, small ornithopods, birds, and smaller theropods such as dromaeosaurs, ornithomimids, and oviraptorosaurs. The herbivorous dinosaurs of the Cretaceous got progressively better at chewing plant matter, and not coincidentally, there was a simultaneous diversification in flowering plants and insects. The nondinosaurian vertebrates of the Cretaceous diversified among this dinosaurian backdrop, and by then, most groups had appeared and were becoming more "modern" in many aspects. This is when crown group mammals greatly diversified too.

Indeed a Stage

These three periods, Triassic, Jurassic, and Cretaceous, comprise the arena in which the early histories of the beast companions played out. As the relatively few dinosaurs went about their business in the Triassic, another set of stories was unfolding all around them. Neither dinosaurs nor nondinosaurs then knew that by the Jurassic, the dinosaurs would be expanding and taking over many roles previously dominated by the other groups we've met in this chapter, while other groups would be continuing on and diversifying significantly themselves. So, by the end of the Late Triassic, many of the groups we'll visit on our journey had taken their place on stage and were ready for the spotlight.

Notes

1. Mesaverde as mentioned here should be in quotation marks, as these rocks are not quite the same age as those at the type section of the Mesaverde Group in southwestern Colorado and New Mexico. There is yet no substitute name for them, however.

2. Conodonts were tiny marine eel-like or wormlike vertebrate predators that lived from the Cambrian to the end of the Triassic.

3. The "clathrate gun hypothesis" of methane release from oceanic and terrestrial sources was developed to help explain periods of rapid warming during the ice age period of the past 2.5 million years, but it has also been suggested to have played a role in the Permian–Triassic crisis.

4. Caseids were bizarre Paleozoic synapsids with the body form of a giant 3 m (10 ft) chuckwalla but a head more like a moderately large, toothed turtle—giant body, tiny head.

A Critical Mass

Invertebrates

2

AMONG THE CONTEMPORARIES of the dinosaurs, we will mostly be visiting the vertebrates in the coming chapters, but there were of course plenty of other animals living at that time. Our interest in this chapter is the animals without backbones, those with shells or exoskeletons or with nothing skeletal at all. Some of the most interesting of these, perhaps not surprisingly, are the insects. The Mesozoic was not a time of giant dragonflies, giant cockroaches, or giant millipedes. Those creatures lived during a particular time in the later part of the preceding Paleozoic era. Unlike the giant sharks and crocodiles that were characteristic of the Miocene epoch of the following Cenozoic than of the Mesozoic, the biggest and most monstrous insects lived before the dinosaurs. But the insects that lived alongside the dinosaurs are revealing, nonetheless.

What's the Buzz?: Insects

A few disclaimers. We've never gotten blood or DNA out of the fossils of Mesozoic insects. Many insects are known from flat compression fossils in pond or lake deposits rather than from amber. True mosquitos, insects that aren't just convergently mosquito-like, appear to have evolved only by the Early Cretaceous. No clones. No "Mr. DNA" (Disappointing, I agree.) But of the menagerie of insects we are, for better or worse, familiar with today, it is remarkable how many were already bugging, or going unnoticed by, the dinosaurs of the Triassic through the Cretaceous.

Some parts of the Mesozoic rock record are nearly devoid of insect fossils. Others are relatively rich in this evidence of past life. In Argentina, the past two decades have brought to light more than 80 species (and hundreds of specimens) of fossil insects just from Triassic-age rocks (fig. 2.1). In North America, however, the main, richly diverse insect deposits are in the Cenozoic, and there are only two truly diverse Mesozoic sites, but they don't match the diversity or preservation of the sites from the Age of Mammals. But Mesozoic insect body fossils are known from many parts of the world and from the Triassic (as we have seen) through the Cretaceous (figs. 2.2 and 2.3). And they survived quite well; today, nearly 75% of named living animal species are insects.

Insects are in the Hexapoda group, generally characterized by features of the jaws, the loss of jointed abdominal appendages, and having 11 abdominal segments primitively. The hexapods evolved from terrestrial arthropods by the Early Devonian period about 410 million years ago, and they are most closely related to the crustaceans. Hexapods and crustaceans

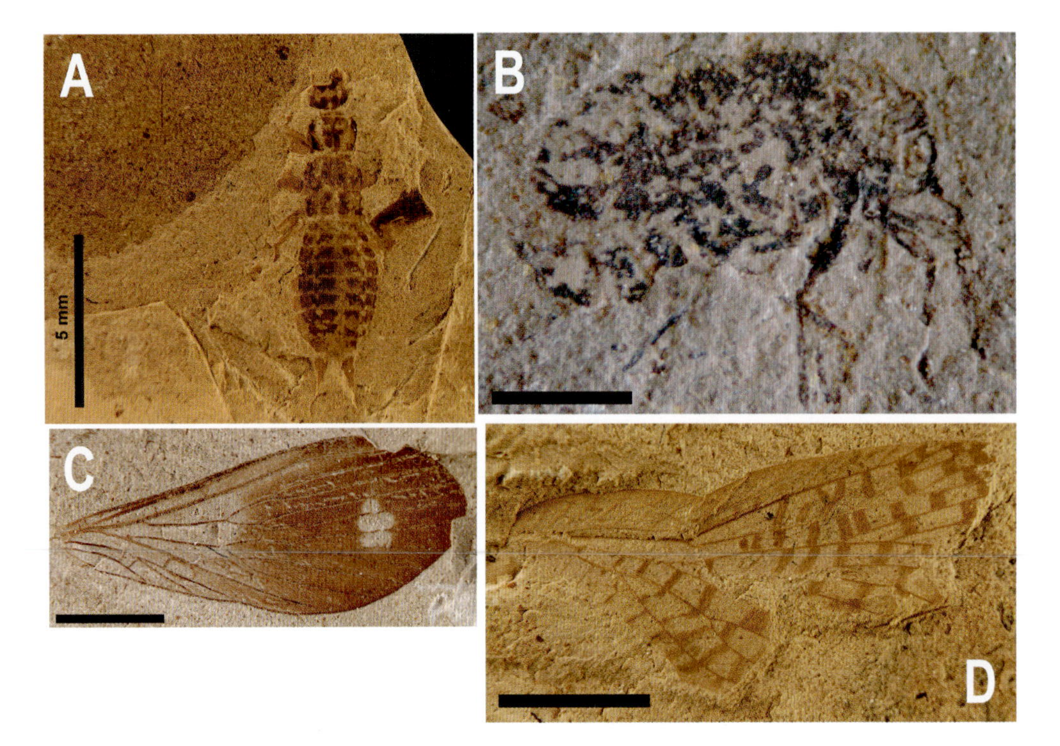

***Above,* 2.1** Some Triassic insects of Argentina. (*A*) Nymph of a relative of dragonflies and damselflies. (*B*) Indeterminate Protopsyllidiidae, a relative of aphids, whiteflies, and scale insects. (*C*) Forewing of the scorpionfly *Duraznochorista* from the Potrerillas Formation. (*D*) Forewing of the basal icebug relative, *Permoshurabia* from the Upper Triassic. Scale bars = 5 mm, except *B*, which = 1 mm. *All images courtesy of Maria Lara.*

***Facing,* 2.2** Some Jurassic insects. (*A*) The Middle Jurassic fly larva *Qiyia* (STMN 65–1) from China, an apparently aquatic ectoparasitic form that may have fed on the blood of salamanders. (*B*) A relative of hairy cicadas, *Shuraboprosbole* from the Middle Jurassic of China (NIGP 149372). (*C*) The froghopper *Jurocercopis* from the Middle Jurassic of China (NIGP 149550). (*D*) An aquatic larva of a caddisfly, in its case (FHPR 11308), from the Upper Jurassic of southeastern Utah, US. (*E*) The beetle *Wuhua* from the Middle Jurassic of China (NIGP 149548). (*F*) The cicadomorph *Talbragarocossus* from the Upper Jurassic of Australia (AM-F.136849), overlapping a fish skeleton. (*G*) The dragonfly *Mesuropetala* from the Upper Jurassic of Germany. Scale bars = 1 cm (*A*, *D*), or in millimeters (*B*, *F*), or = 2 mm (*C*), or = 3 mm (*E*). *All photos courtesy of Bo Wang and Maria Lara, except D, which is courtesy of Tom Howells, and G, by author.*

(Pancrustacea) are allied with the myriapods (centipedes and millipedes) in the Mandibulata.

Arthropods had made their way onto land from the marine and estuarine realms possibly by the Ordovician but at least by the Silurian period (470–420 mya), and the fact that all the earliest terrestrial arthropods were predatory suggests that they may have first been drawn to otherwise empty dry land as a refuge from predators during the vulnerable parts of their reproductive cycles. Why stick around and get attacked when you can escape to a less crowded environment? Such a draw also proved irresistible to early tetrapods (at least in part) and flying fish at various times. That there

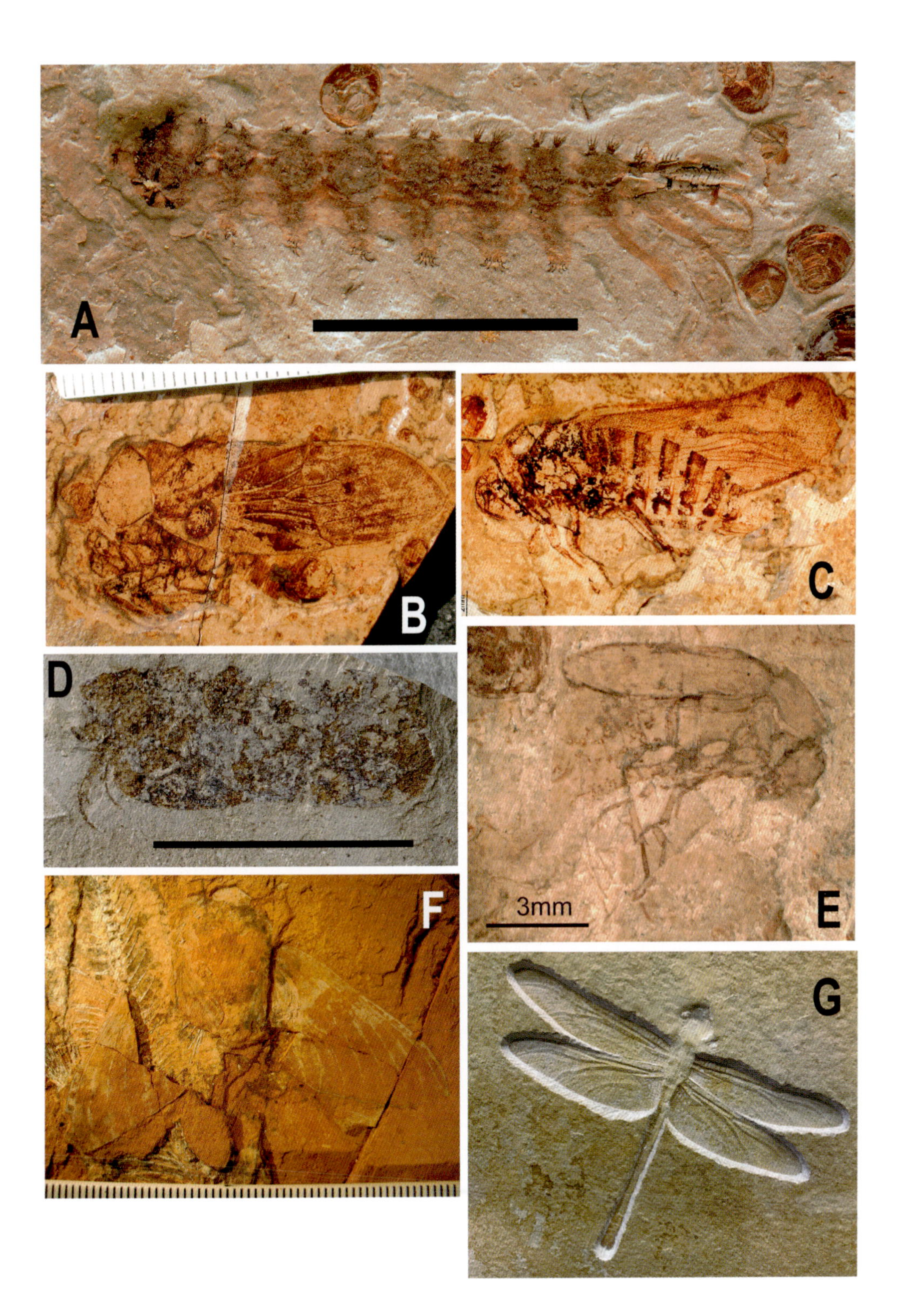
A
B
C
D
E
3mm
F
G

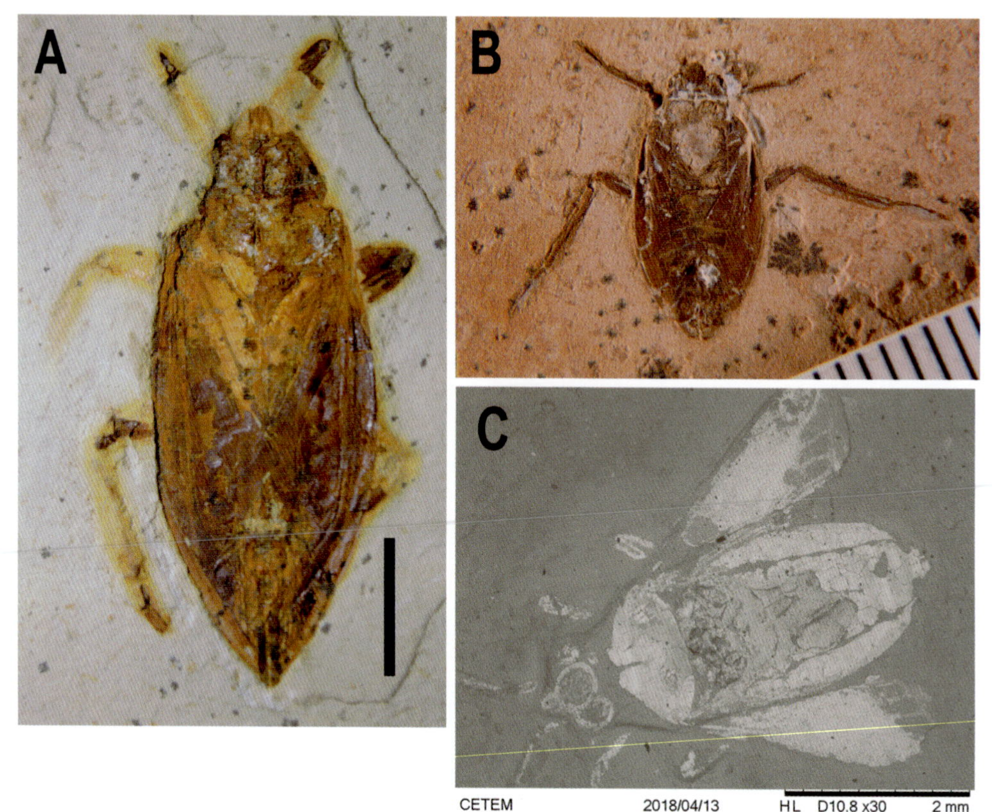

2.3 Some Cretaceous insects of Brazil. (*A*) The giant water bug *Lethocerus placidus* from the Early Cretaceous–age Crato Formation (MPSC I217). (*B*) Giant water bug *Neponymphes*, also from the Crato Formation (MCT 6955). (*C*) Shore bug *Olindasalda*, Crato Formation (MCT 6959). Scale bar in *A* = 1 cm, scale in *B* in mm, scale in *C* = 2 mm. *All images courtesy of Dionizio Moura-Júnior and Maria Lara.*

appear to be marine stem[1] hexapods suggests that the transition to land by the insects' ancestors was a separate process from that undertaken by the myriapods; in fact, the transition to land likely occurred at least six times in arthropods and their relatives, separately in hexapods, myriapods, isopods (crustaceans), chelicerates (spiders and scorpions), tardigrades (water bears), and onychophorans (velvet worms). The terrestrial origin of insects also indicates that mayflies, dragonflies, and water bugs are secondarily aquatic.

Insects are characterized very broadly by the presence of six legs in adults of most species and, in many cases, by wings, the latter of which evolved by the late Carboniferous (310 mya) from modified extensions of the thoracic terga (similar to those in silverfish) or from modified gills similar to the abdominal gills of young mayflies. At the Permian–Triassic boundary, insects suffered the extinction of a full seven orders of their Paleozoic representatives, and an eighth continued only a little way into the Mesozoic. Seven more orders originated in the Paleozoic and are still around today, five of today's orders originated in the Mesozoic, and one existed only in the early Mesozoic. These numbers and the family-level diversity in each order indicate that the Mesozoic to Recent insect fauna is rather different from that which existed in the Paleozoic. After the major extinction of insects at the Permian–Triassic boundary, the net diversification of insect families (origination rate minus extinction rate) has been fairly steady and just on the plus side—indicating very gradual diversity growth. In fact, insect family richness appears to have been more or less steady since the Early Cretaceous.

Some of the most productive localities for Mesozoic fossil insects are in Virginia and Australia for the Triassic; Kazakhstan and Germany for the Jurassic; and New Jersey, Brazil, Lebanon, and Myanmar for the Cretaceous. The New Jersey, Lebanon, and Myanmar localities preserve the insects in three dimensions in amber pieces, while the others are more typical compression fossils in fine-grained siliciclastic rocks. Together, these sites and a number of other small ones produce most of the known diversity of Mesozoic insects.

So, what insects were around to torment the dinosaurs, or go barely noticed by them, during the Mesozoic? Almost every insect type we know today. I'm currently sitting in a meadow in the Snowy Mountains of Wyoming, and buzzing or crawling around me on this rather average (occasionally breezy) July morning at 2,621 m (8,600 ft) altitude are butterflies, moths, beetles, ants, mosquitos, and an unhealthy helping of flies (well, at least they are meadow flies, not dumpster flies, but there are a *lot* of them). There are also spiders within view, which are not insects, of course, but arachnids. Down on the banks of a nearby river are adult mayflies, with dinosaur-food horsetail plants scattered around under spruces and firs. In the swift current of the river are the larvae of the mayflies, primordial and almost Cambrian-looking, clinging to the undersides of the cobbles on the bottom of the stream (along with some small gastropods—more on them later).

Even out on the playa, the barren, dry lake beds of the Great Basin in the western United States, one encounters ants, grasshoppers, tarantula hawks, beetles, and scorpions (chelicerates), among others. Then, in the right spots, after sunset, the flies and other flying insects attract thrillingly low- and slow-flying bats that gently buzz your head like fuzzy mammalian butterflies. The insects of the Mesozoic probably drove much of the terrestrial ecosystem functions then as well.

All of these insect and arachnid groups were contemporaries of the dinosaurs for some or all of the Mesozoic. Butterflies and moths appear in the Late Jurassic, with forms known from Siberia during that epoch, plus Cretaceous occurrences from Brazil, Lebanon, and Myanmar. Beetles occur in the Triassic of Virginia and Australia; the Late Jurassic of Germany, North America, and Kazakhstan; and the Cretaceous of Spain, Siberia, Brazil, Myanmar, and New Jersey. Among these occurrences are scarabs, dermestids, and weevils, among others. Ants are known from at least the Cretaceous of New Jersey, Asia, and Brazil. The Las Hoyas deposit, in the Lower Cretaceous of Spain, has produced wasps and scorpionflies. True mosquitos probably evolved by the Early Cretaceous, with forms known from amber in Canada and Myanmar. Other true flies such as craneflies, midges, fungus gnats, and horseflies are known from the Triassic through the Cretaceous at a number of localities. And mayflies occurred throughout the Mesozoic, with specimens known from (among others) the Late Jurassic of Germany and the Cretaceous of Brazil, New Jersey, and Siberia. Spider wasps, related to tarantula hawks, have been found in the Cretaceous of Myanmar. Other types of insects known from Mesozoic rocks include dragonflies and damselflies, webspinners, stoneflies, water striders, zorapterans, orthopterans (crickets, grasshoppers, katydids, locusts), stick insects, titanopterans, earwigs, ice crawlers, roaches (or "roachoids"), termites, mantises, bark lice and true lice, thrips, primitive hemipterans, whiteflies, aphids, scale insects, cicadas, true bugs, snakeflies, lacewings, antlions, strepsipterans (parasites), wasps and bees, scorpionflies, fleas, and caddisflies. The Mesozoic was likely nearly as infested with insects as the world is today. Bees, however, didn't appear until the latest Cretaceous (in New Jersey amber), but they appeared morphologically and socially derived, so their true origins are probably back in the Early Cretaceous around the same time as the origin of early flowering plants. And as noted, mosquitos probably didn't appear until the Early Cretaceous either. Whatever bit the Jurassic dinosaurs of *Jurassic Park*, it probably was not a true mosquito.

The chronological association of social insects like bees with the first flowering plants may not be a coincidence. The coevolution of insects that utilize nectar and spread pollen among the flowers that produce these temptations probably fed the diversification of both during the Cretaceous.

The noninsect centipedes and millipedes (Myriapoda) are also known from fossils in Mesozoic rocks, mainly from numerous sites in Brazil, Europe, and Asia. Myriapods appear to have originated early in the Paleozoic and are grouped with insects and crustaceans in a large group known as the Mandibulata.

Southeastern Utah is home to the outdoor wipeout wonderland of Moab, sandwiched between Arches and Canyonlands National Parks, with the La Sal Mountains to the east. The emergency room at Moab's nice new hospital has a regional map called the "biff board" where patients can use colored pins to locate where in the outside playground of the canyon country around the town they ate it on their mountain bike, fell from climbing ropes, crashed while BASE jumping, flipped their raft, or rolled their ATV, among many other forms of personal disaster that people stumble into while playing in the wilderness in the area. The town of 5000 residents is besieged by some two million visitors per year, so the tiny hospital gets more than its share of backcountry emergency victims. Moab is a busy place almost year-round, but the country around it is only a little less so in the winter. Surprisingly for a town in the desert, Moab has its share of gnats (biting midges) and mosquitos. I learned this the hard way when I did a week of fieldwork in the Morrison Formation in June after a very wet winter. I ended up duct-taping my sleeves and pant legs shut at the wrists and ankles. But this part of Utah doesn't just have modern insects. Fossils of Mesozoic insects are beginning to show up as well.

About an hour and a half south of Moab is Blanding (still in Utah), a much quieter town with a slightly less busy backcountry that is not far from the landscape beauty of Bears Ears and Natural Bridges National Monuments and the ancient Ancestral Puebloan settlement of Hovenweep (also a national monument). Just outside Blanding is a locality in the Morrison Formation that has produced hundreds of plant fossils. One day just a couple years ago, I was digging there and turned up a weird plant—I thought. After a couple of seasons there (the site had only been found in late 2016), my brain was already focused on ginkgoes, ferns, and *Czekanowskia*, another type of ginkgophyte, so I didn't recognize that the weird plant I couldn't identify was in fact an insect and not a plant at all. I believe some of my mind'sinattentiveness may be forgiven, however, because in 140 years of digging in the Morrison Formation, paleontologists had unearthed a grand total of one insect body fossil from the unit before this (an orthopteran from Temple Canyon in Colorado only described in 2011). An insect simply wasn't the search image one had at a site known only for plants and in a formation known for seemingly everything *but* insects. It wasn't until a couple years later at my office, when I put this mystery specimen under a microscope to try to figure out what kind of plant it was, that I finally noticed that the veins I was seeing in the specimen were the veins of a wing and not of a leaf! And it was larger than we had expected. María Lara, a paleoentomologist in Argentina, recognized that this second insect fossil from the Morrison Formation was a nepomorph hemipteran (among the true bugs) and possibly a relative of today's giant water bugs (belostomatids), large semiaquatic predators with a sharp mouth proboscis that can hunt even vertebrates such as frogs, fish, and salamanders. These nepomorphs are known from the Triassic of North America and the Cretaceous of Brazil, but this (now named *Morrisonnepa*) was the first that had been found in North America in the Late Jurassic.

The same site has also produced fossils of larval caddisfly cases, and these cases (minus the larva itself) have also been found at the Fruita Paleontological Area and Temple Canyon, Colorado, in the Morrison Formation. Caddisflies are aquatic animals as larvae, and they build themselves a protective tube to live in, with head and legs sticking out (or just inside) one end. In the Morrison Formation, these cases have been found to have been made by the larvae by sticking together numerous sand grains, arthropod shell fragments, or fecal pellets, whatever is available. After their larval stage, caddisflies shed the case and become terrestrial fliers for a brief period, but modern species exhibit a wide range of feeding ecologies as aquatic larvae. Most recently, a beetle elytron turned up in the lake deposit in the Morrison Formation in Utah too.

Insect–plant interactions are a large area of study for paleoentomologists and paleobotanists too. Fossil leaves from all Mesozoic periods sometimes demonstrate damage from herbivorous insects feeding on the plants. Although we have yet to identify any of this damage from the Jurassic site near Blanding, Cretaceous leaves from many sites seem to have these features a little more commonly. As an example of the diversity seen in these fossils, one Late Triassic formation in South Africa produces leaves with at least 44 different damage types preserved among the flora. Clearly, the herbivorous insects of the Mesozoic had long since found, and were successfully exploiting, the plant food sources of the time.

Chelicerate Nightmares

Insects are just one group of arthropods from the Mesozoic. Among the chelicerates, dinosaur ecosystems sported scorpions, spiders, ticks, and, in the seas, horseshoe crabs. Among the arachnids, spiders appeared during the Carboniferous (about 300 million years ago) in Europe and North America, and these first spiders likely were, similar to modern forms, terrestrial predators capable of silk spinning, although their spinnerets were not at the posterior end of the abdomen at that time. By the Early Triassic, forms with the spinneret at the end of the abdomen had appeared in France, and the first spiders related to modern orb weavers appeared by the Jurassic, with cobwebs and probably classic spiral wheel–shaped webs being spun by these species by the Early Cretaceous. Amber found in the Early Cretaceous of Spain, in fact, preserves an actual spider web with several insects and arachnids trapped in it (wasp, beetle, and mite).

Among other arachnid fossils known from the Age of the Hyped Reptiles (okay, they've earned some of the hype), there were also microwhip scorpions (palpigrades) in the Late Jurassic of Germany, ticks in the Cretaceous of Asia and North America,[2] scorpions in the Triassic of North America and Europe and the Cretaceous of Brazil, and mites in the Cretaceous of Europe. These groups all seem to have originated in the Paleozoic and to have lived in the terrestrial environments of the dinosaurs and their contemporaries throughout the Mesozoic. Considering the presence of mites during the Age of Dinosaurs and the fact that so many nonavian dinosaurs appear to have been feathered, it's hard not to imagine some dinosaurs dust-bathing like modern chickens to try to keep mites under control. In fact, there may be some trace fossil evidence of this in the Cretaceous of western Colorado.[3]

One fossil type common to many Mesozoic formations among arthropods is the xiphosurids, or horseshoe crabs. These chelicerates, which are familiar on the beaches of the East Coast and Gulf Coast of North America and Asia today,[4] actually first appeared way back in the Ordovician Period, alongside trilobites, about 445 million years ago. This was *Lunataspis*, the "crescent moon shield" described recently from rocks in Manitoba, Canada. It was relatively primitive, but it was recognizable as a horseshoe crab by anyone who knows today's species. From this Ordovician point on, horseshoe crabs were in parts of many shallow, sandy to muddy marine and tidal settings for the rest of geological time (so far), feeding on bottom-dwelling mollusks, crustaceans, and worms. In the Mesozoic, they are known from many deposits, including as particularly well-preserved specimens in the Late Jurassic lagoons that existed in what is now Germany. Their tracks are common in some deposits too. The Circle Cliffs uplift, west of Capitol Reef National Park in southern Utah, contains at its center outcrops of the Moenkopi Formation, a unit of the Early–Middle Triassic age that contains plated beds of brick-red shale, many with trackways of horseshoe crabs (trace fossils known as *Kouphichnium*) and fin traces of fish (*Undichna*). When we were with the Utah Geological Survey years ago, Josh Smith and I happened across this site during a paleo survey of Grand Staircase-Escalante National Monument. We spent several days pulling out slabs crossed by the trackways of horseshoe crabs, the tracks consisting of widely spaced, paired little fleur-de-lis-shaped indentations in the shale, although they were upside down relative to the direction of travel. These trace fossils in the Moenkopi occur in deposits representing what were likely shallow marine tidal flats, and similar traces are known from many other formations throughout the Mesozoic. In addition to the Moenkopi in the Middle Triassic of the southwestern United States, trackways indicate the presence of horseshoe crabs in the Early Triassic of China and Spain; the Middle Triassic of Europe and Idaho; the Late Triassic of Arizona; the Early Jurassic of Utah; the Middle Jurassic of Morocco and the UK; the Late Jurassic of France, Germany, and India; the Early Cretaceous of Japan and Argentina; and the Late Cretaceous of Mexico, Montana, and Kansas—as just a few examples. Obviously, horseshoe crabs inhabited waters around the dinosaurs in many places and for the whole of the Mesozoic—and they had ecologically similar roles to the horseshoe crabs of today.

Menu Items: Crustaceans

The tastiest invertebrate group during the Age of Dinosaurs was that containing the lobsters, shrimp, and crabs, Crustacea.[5] Arising in stem taxa during the Middle Cambrian, Crustacea today includes lobsters; crabs; and their cousins, the shrimp, ostracods, barnacles, and crayfish. Isopods (including "pill bugs," or "rollie pollies") are crustaceans too. All of these groups inhabited the world of the dinosaurs, although only the crayfish, ostracods, and (by the Cretaceous) some isopods were freshwater species (to terrestrial among isopods). Amazingly, there is a Late Jurassic–age lagoon deposit in Germany that has yielded most of these groups: marine isopods, shrimp, lobsters, ostracods, and barnacles. In the hill country of southeastern France, there are fossil shrimp and lobsters in Middle Jurassic rocks deposited in the ancient western Tethys Sea. Fossil crabs are present in a number of the shallow marine and estuary deposits of the epeiric seas of the Cretaceous around the world, including a particular diversity in the mangrove swamp and estuary deposits in Egypt. Ostracods are tiny arthropods that live in the water column in ponds, lakes, and oceans, tucked within a pair of clam-shaped shells; these animals have been used to attempt to correlate rocks between field areas and can be common in some formations. Isopod fossils are quite rare even in marine deposits. But possibly most photogenic among freshwater crustaceans are the crayfish. These relatives of marine lobsters are known from only a couple dozen fossil localities, although they can be quite abundant at some individual sites.

There are relatively few fossil crayfish deposits worldwide, with only ten species described so far. Crayfish probably evolved from a now-extinct lineage of lobsters in the late Paleozoic or early Mesozoic, but their fossil record is rather spotty prior to 30 million years ago. Some particularly well-preserved specimens are known from lake deposits in the Lower Cretaceous of China and in wetlands deposits of about the same age at Las Hoyas, Spain, and they occur in several other isolated spots. Of course, crayfish survive today as well, and sometimes in great numbers!

Go to just about any major stream in the southeastern United States today, and you'll eventually see crayfish among the rocks in the shallows. For that matter, plenty of streams elsewhere in North America (Central America up to southern Canada, although they were probably only originally native to the south and east) are crayfish habitats today. Europe, eastern China, New Zealand, New Guinea, Australia, and parts of South America, Africa, and Madagascar also contain crayfish. These freshwater relatives of lobsters have lived in rivers and lakes of Earth's ecosystems for hundreds of millions of years, and they were probably nearly as common during the Age of Dinosaurs.

The Mygatt-Moore Quarry is a site in western Colorado less than two miles from the Utah border. In sight of Interstate 70, if you know where to look, the quarry is in the Morrison Formation and is about 152 million years old, the bones and mud preserved in place from a spot on the Late Jurassic floodplain where carcasses of large herbivores were scavenged and their bones gouged by predators such as *Allosaurus* and *Ceratosaurus*. The bones were also marked by insects and other invertebrates as the remaining flesh was stripped from the bones. But after this gladiator-pit-of-scavenging's time was up, the low point in the floodplain it had occupied became a small, perennial lake, complete with fish, snails, downed tree branches, ripple marks caused by wave action—and crayfish. A layer just above the quarry has produced a number of fish skeletons over the years, but rarer, a nearly complete crayfish exoskeleton was found in that layer in the late 1990s (fig. 2.4*B*). This crayfish fossil has been studied by Steve Hasiotis and Jim Kirkland and is due to be described again soon. A second rare crayfish fossil out of the Morrison Formation was recently found in southeastern Utah (fig. 2.4A) at the same site that produced the insects and plants mentioned earlier.

These Late Jurassic fossil crayfish and their better-preserved counterparts in China show animals of similar size and with similar morphology to an average crayfish today, basically a miniature freshwater lobster. Crayfish, lobsters, and crabs are all decapods, meaning they have "ten feet"; in this case, four pairs of walking legs plus a grasping appendage on each side of the body. On my desk, I have a modern member of the crayfish genus *Faxonius*, preserved in alcohol and found as a carcass in the Green River last summer. The body is about 10 cm (4 in) long and is divided into a head and thorax protected by a cephalothoracic shield; an abdomen with about a half dozen segments, each with a reduced, nonwalking leg (pleopod); and a telson and four uropods splayed transversely in a "fan." The walking legs (periopods) and grasping appendages (cheliped) extend from under the thorax. Fossil crayfish of the Mesozoic from Asia and North America are very

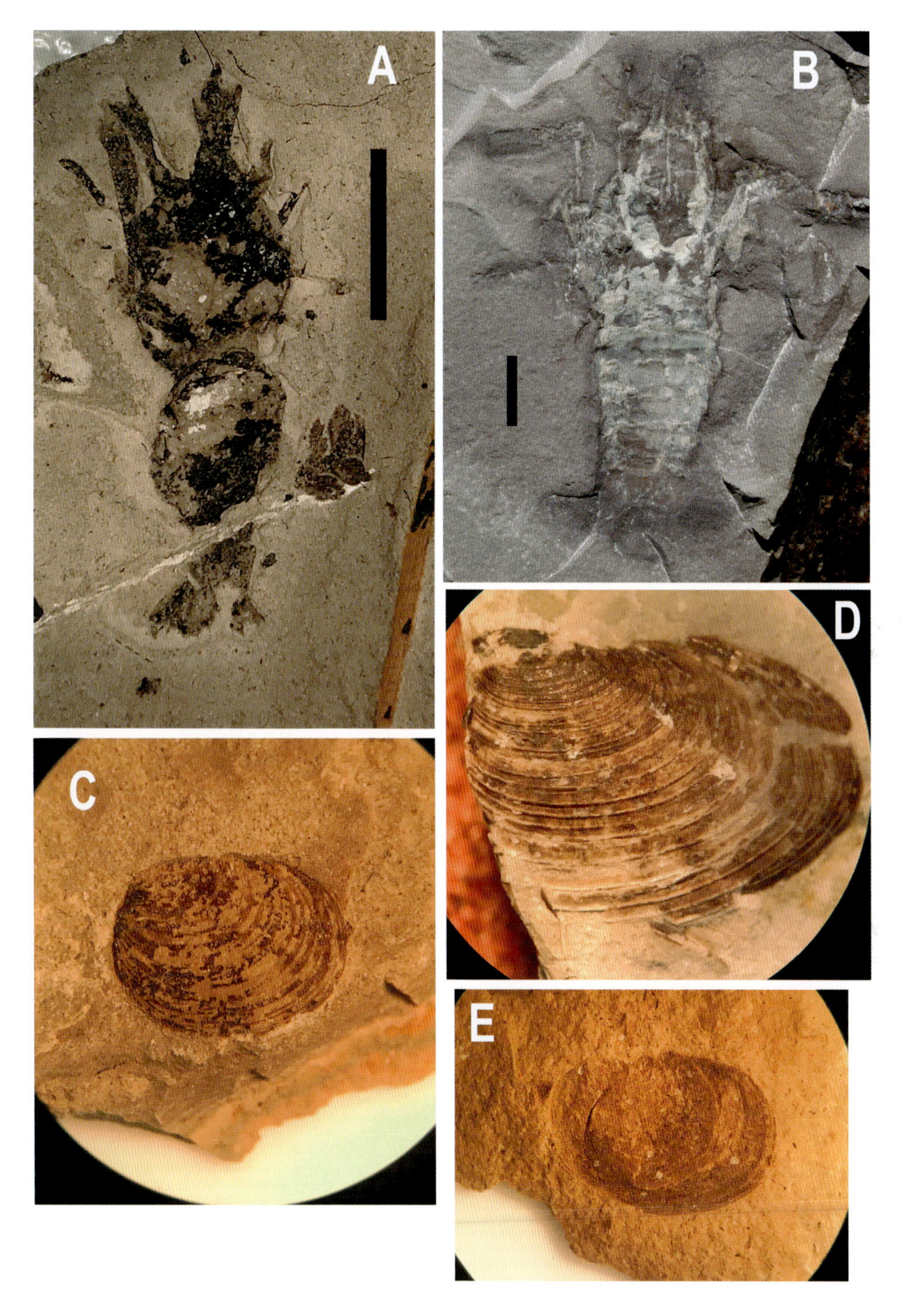

2.4 Some freshwater arthropods of the Mesozoic, here as demonstrated by the Late Jurassic–age Morrison Formation of North America. (*A*, *B*) Crayfish from pond deposits in southeastern Utah (FHPR 11301) (*A*) and western Colorado (specimen located at Museums of Western Colorado MWC 5443) (*B*). (*C–E*) Diplostracan ("clam shrimp," or conchostracan) shells from southeastern Utah (FHPR 17128) (*C*), eastern Utah (FHPR 16972) (*D*), and central Colorado (FHPR 16967) (*E*). Scale bars in *A* and *B* = 1 cm. Microscope view diameter in *C–E* = 15 mm. *Photos A and B courtesy of Tom Howells; C–E by author.*

similar to this form. Add to this the eyes and antennae that are sometimes also preserved, and there you have it: the diminutive freshwater lobster that lived in lakes and rivers of the dinosaur age.

In addition to the Morrison's two as yet unnamed crayfish, there are *Palaeocambarus licenti* and *Cricoidoscelosus aethus* from the Early Cretaceous of Mongolia and China. Then there are *Austropotamobius* and *Protastacus* from the Early Cretaceous of Europe. And that's about it for Mesozoic crayfish fossils. Freshwater pond deposits are rare, and the animals are small, delicate, and prone to damage before burial and fossilization. The crayfish out of the Morrison Formation are only about 4–6 cm (~2 in) long, but some of the other fossil taxa get a bit bigger, up to about the size of the modern one I described above. Because their exoskeletons are far less durable than bones or shells, they are far less common than vertebrates or mollusks.

The oldest fossils are traces from the late Paleozoic, and rare body fossils, like the ones from the Morrison and Asia and Europe, are known from the middle Mesozoic. Modern or ancient, most crayfish live under rocks in streams and lakes eating a generally omnivorous diet. In turn, they may well have been fed on by lungfish and small crocodiles. Ancient crayfish also appear to have burrowed, and their fossil burrows are found commonly, especially in the Chinle and Morrison Formations.

Clam shrimp (also known as diplostracans, spinicaudatans, or conchostracans) are common in freshwater deposits of the Mesozoic (fig. 2.4C–2.4*E*). These arthropods look more or less like brine shrimp with a pair of clam shells draped over their back; we generally find only the shells, but they are common enough that they probably lived in many of the same ponds and lakes on the dinosaurian landscapes of the Mesozoic as snails and clams (which we will see later).

Worms of a Different Color: Annelids

Annelid worms today are typified by the terrestrial earthworm of so many a backyard garden, by freshwater leeches, and by the bristly marine polychaete worms. All soft parts, annelids are rare as fossils, but a few do occur. In the Upper Jurassic shallow marine deposits of Germany, there are two polychaete annelid genera, plus a nonpolychaete that lived on driftwood and ammonoid shells.

Dinosaurs and other vertebrates of continental landscapes may have encountered annelids as well. Freshwater deposits dating from the Early Jurassic of Australia have produced egg cocoons of probable leeches, and these types of fossils have also been reported from the Late Triassic. So, if dinosaurs were capable of being disgusted, they might have been so by Mesozoic leeches if they waded through the wrong waters.

Octopus's Garden: Mollusks

Around the Parthenon, on the rough plateau atop the commanding hill known as the Acropolis, in Athens, Greece, is a low outcrop of gray limestone, early Late Cretaceous in age, which has been thrust over younger Cretaceous rocks. It is the basement on which the ancient collection of buildings was constructed. The marble of those buildings is composed of early Mesozoic limestone metamorphosed into marble during the Late Jurassic. But the foundation stones of the Parthenon and other buildings on the Acropolis are blocks of limestone from the Athens area, and these blocks contain anywhere from several to many gastropod and bivalve fossils that may be Mesozoic in age too. These are visible right in the blocks while standing at the base of the Parthenon. Mollusks are so common as fossils in stones that are sought after for building that they appear in many structures around the world.

The mollusks as a group include gastropods (snails of freshwater, marine, and terrestrial environments), bivalves (freshwater and marine clams, oysters, and their relatives), cephalopods (squids, octopus, and nautilus), the multiplated marine chitons, and a few others that are relatively rare. Gastropods and bivalves appeared at least by the earliest Cambrian (~540 Ma) as part of the "small shelly faunas" of the pretrilobitic Early Cambrian diversification that started at the very end of the Precambrian. Representatives of the groups at the time were tiny, often only about 1 mm (0.04 in) in diameter, and they were rare elements of faunas for millions of years, but with the rise of the Paleozoic Evolutionary Fauna in the Ordovician, they became larger and more common. By the Mesozoic, they dominated the marine faunas. Of course, freshwater bivalves and gastropods evolved as well. Today, these mollusk groups include everything from conches and land snails to giant clams, mussels, and scallops.

Gastropods and bivalves of the Mesozoic are almost ubiquitous in both freshwater and marine deposits. It's almost hard *not* to find snails in dinosaur deposits and their offshore counterparts from the Triassic through the Cretaceous. Most freshwater deposits that one encounters in the field, ancient lakes, ponds, and rivers have a least one species of gastropod, often several, just in the western United States, from the Chinle Formation up through the Morrison (fig. 2.5*B*) and into numerous Cretaceous deposits. Bivalves are rather similar in that they are nearly everywhere; in the Morrison Formation, fossils of unionid bivalves (fig. 2.5C) are found in river and pond deposits, which indicates year-round water because the larvae of these clams go through a stage when they embed themselves in the gills of ray-finned fish that inhabit bodies of water that don't dry up. So freshwater clams, of all things, can give you clues about the paleoenvironments in which dinosaurs are found.

Then there are the marine bivalves (fig. 2.5*D* and 2.5*E*) and gastropods of the Mesozoic. These are more numerous and more species-rich than their freshwater counterparts. And, in the case of *Platyceramus*, they are huge. In eastern Utah, where driving Interstate 70 east or west reminds one of crossing a moonscape, there are several localities in the Mancos Shale with giant marine inocermid bivalves of the genus *Platyceramus*. This mussel-like relative of clams is commonly up to 1 m (3 ft) across the shells but can get significantly larger, up to more than 2 m (6 ft). Fragments of these giant shells are commonly scattered over the ground at these sites, and even these partial specimens demonstrate that the outside of the animals' shells served

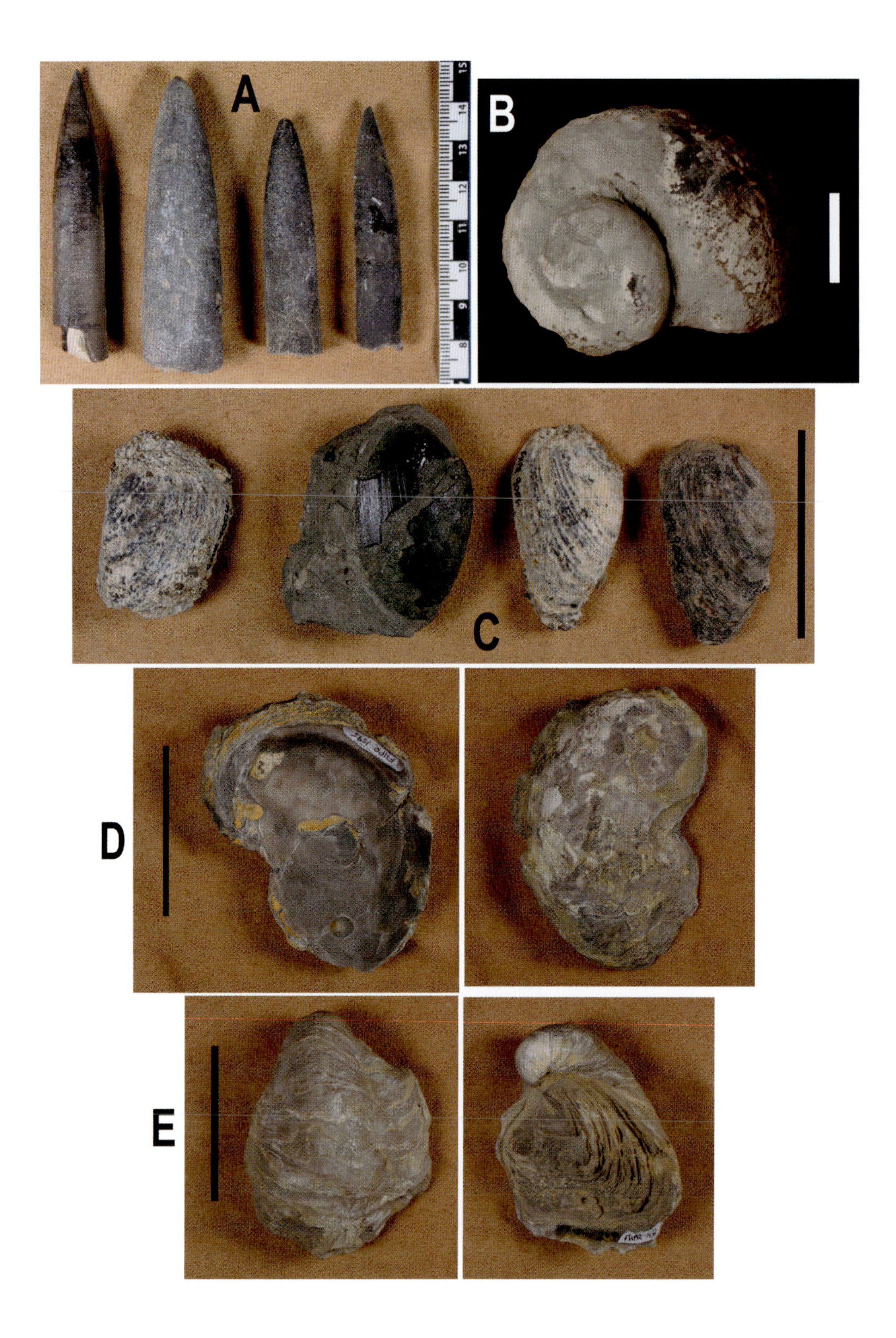
A
B
C
D
E

***Facing*, 2.5** Some mollusk groups of the Mesozoic. (*A*) Belemnites (*Pachyteuthis*) from the Middle–Upper Jurassic Stump Formation of Utah (FHPR 1957, 2005, 1960, 2010). Fossils of these squid-like animals are common in shallow continental seaway deposits of the Rocky Mountain region. (*B*) Cast of the freshwater snail *Pila* from the Upper Jurassic Morrison Formation of northeastern Wyoming (FHPR 2081). (*C*) Freshwater unionid clams from the Upper Jurassic Morrison Formation of Utah (FHPR 9006 and 1581). These mollusks depend on the gills of ray-finned fish during their larval stages. (*D*) Two views of the bivalve *Ostrea*? from the Upper Cretaceous Austin Chalk of Texas (FHPR 1595). (*E*) Two views of the oyster *Gryphea* from the Lower Cretaceous Edwards Limestone of Texas. All photos by author. Scale in *A* in cm, scale in *B* = 1 cm (0.4 in), scales in *C–E* = 5 cm (2 in).

as attachment sites for other sessile animals, as the surfaces are commonly covered with shells of oysters as well as juvenile *Platyceramus* and barnacles. The seas were a story in themselves during the Mesozoic.

Cephalopod mollusks were quite diverse and abundant during the Mesozoic, but most of that richness was in the now-extinct ammonoids. The cephalopods we are familiar with today are rarer as fossils due in part to preservation—they don't have many (or any) hard parts, unlike the ammonoids. During the Middle Jurassic, in the seas of Europe, there was *Proteroctopus*, a basal vampyropod relative of octopus and vampire squid, found now in the La Voulte lagerstätte (fossil bed with exceptional preservation) south of Lyon in southeastern France. Octopus we all know and love as the worldwide, sometimes trouble-making cephalopods of surprising intelligence, the highest known for invertebrates. Octopus in fact are active predators and can camouflage, communicate, learn, use tools, and problem solve ("they remember . . ."). Vampire squid (a single species) are deep-living (600–900 m [2,000–3,000 ft]), close relatives of octopus that are adapted to low-oxygen conditions. They grow up to 30 cm (1 ft) long and feed mostly on detritus composed mostly of zooplankton. The ancient *Proteroctopus* had a short, bulbous head with eyes and likely a siphon (or funnel, for jet propulsion), and two lateral fins, plus eight arms, each with a double row of suckers. It was only about 15 cm (6 in) long but was likely predatory. Also found in the same deposits in France are *Vampyronassa*, a more derived relative of the vampire squid, and *Gramadella*, a squid.

The Jurassic marine deposits of the western United States, units such as the Sundance and Stump formations, often contain abundant fossils of the belemnoid cephalopod *Pachyteuthis* (fig. 2.5A). The fossil is an element known as the guard, which is shaped like a cigar or bullet, though the animal it was part of was similar in form to a modern squid.

On the Normandy coast of France, northeast of the village of Bayeaux (home of the tapestry depicting the AD 1066 Norman invasion of England) and northwest of Caen, is the small town of Arromanches. This was the center of a Mulberry harbor used by Allied forces after the June 1944 D-Day landings. The town still sports displays of a tank and a landing craft of the period, and just offshore are the remains of several Phoenix breakwaters from the old harbor. The open sand of the shallowly sloping beach is often covered with sandworm casts during low tide, and along the beach just east of town, the cliffs to the south are composed of limestone and shale of Jurassic age, with benches exposed just above the high tide line between the sand and cliffs. This is the area of Gold Beach during the D-Day invasion, where the 50th Division of the Second British Army landed and eventually captured the town back from Wehrmacht forces by the end of June 6. Nestled in the limestones of these layers at the base of those cliffs, though, are fossils, including those of ammonoids, relatives of modern squids and nautiluses that were so common in the Mesozoic seas. I stumbled across the imprint of one of these ammonoids there years ago, a specimen that was about 25 cm (10 in) in diameter, and I couldn't help but wonder if it had been exposed to the chaos of that day in 1944, positioned as it was just yards from what had been the area of the right flank of the British 50th.

Ammonoids appeared during the Paleozoic and lasted throughout the Mesozoic until the end of the Age of Dinosaurs. They were contemporaries of the dinosaurs during their reign but because they lived in the oceans, they were ecologically separate from the giant reptiles. The ammonoids were related to squids and nautilus but were their own group. Their fossils are common in marine deposits from the Triassic through the end of the Cretaceous, and in some deposits, they are particularly abundant and diverse (fig. 2.6). They were also worldwide in their distribution, so their abundance and diversity are common in ancient marine deposits just about everywhere. Because of the diversity and short duration of the species and their vast distribution, ammonoid fossils have been used extensively for biostratigraphic correlation of marine rocks.

I've sometimes described ammonoids as "squid in a snail shell," although they are not exactly squids (just squid-like animals) and their shells generally coil in the same plane (unlike a majority of snails). A chambered nautilus is a pretty good model of the general idea of an ammonoid—a coiled shell with the open end occupied by an animal with tentacles sticking out and eyes at the base. They could probably regulate their depth and swim backward by jet propulsion. The sutures within the shell, the contacts between

***Facing,* 2.6** Ammonoids of the Mesozoic, just a few of them. (*A*) A large ammonite (next to rock hammer) as found in the Mancos Shale of Utah. (*B*) *Prionocyclus* from the Juana Lopez Member of the Mancos. (*C*) *Metoicoceras*, in side and end view, from the Tropic Shale of Utah. Scale in centimeters. (*D*) Tiny *Collignoniceras* from the Mancos Shale near Moab, Utah (FHPR 16862). View diameter is 15 mm (0.6 in). (*E*) Cretaceous ammonite from the Goodland Limestone of Texas (FHPR 1776). Scale bar = 5 cm. (*F*) Numerous *Dactylioceras* from the Early Jurassic of Germany (DMNH 5099). (*G*) Cretaceous *Tragodesmoceras* from Copper Canyon, Mexico (FHPR 13679). (*H*) Cretaceous *Hoploscaphites* (FHPR 1779). (*I*) *Aspenites* from the Triassic-age Thaynes Formation of Utah. Scale in centimeters (numbered). (*J*) The Cretaceous baculite *Sciponoceras*. Scale in centimeters. *Images in A–C, H, I, and J courtesy of Kevin Bylund and www.ammonoid.com; images in D–H by author.*

A
B
C
D
E
F
G
H
I
J

the walls of each new living chamber and the outside shell surface, could be incredibly, artistically, and fractally complex, and in fact, these sutures help identify different species in many genera. The shapes, coiling, spines, ridges, and occasional spiring of the overall shell morphology were so incredibly diverse during the Mesozoic as to seem almost limitless. Of course, there is a finite number of species of ammonoids known, give or take 10,000 in fact, but that's about as diverse as dinosaurs—I mean, birds—are today. So fossils of the cephalopod group Ammonoidea are amazingly common, are almost artistic pieces of biological work, and as biostratigraphic tools are some of the most useful fossils around. They are, in many cases, how we can tell exactly what age certain dinosaur deposits are; if marine and freshwater deposits are interbedded, we can tell from ammonoids in the marine beds how old the dinosaurs are and to what other units they correlate around the world far more precisely than we can from almost any terrestrial or freshwater fossils in the dinosaur beds. And, because many significant marine units with many ammonoids (e.g., the Pierre Shale in the Great Plains of North America) also have a number of ash beds in them, we can correlate these ammonoid zones with particular numerical ages, again allowing us to date dinosaur ecosystems. This is the kind of precision that the dinosaurs themselves haven't been willing to give up—or that we at least haven't been able to tease out of them. So the ammonoids are our friends.

Ammonoids are beautiful fossils to collect just for fun as well. Their morphologies are so fascinatingly complex and striking that I have a hard time calling them humble, but they are helpful to us to a degree far out of proportion to their anonymity among the general public. Dig up an ammonoid fossil today and tell it "thanks."

Unfortunately, ammonoids disappeared along with dinosaurs at the end of the Cretaceous. But they had a darn spectacular run while they lasted. And their legacy is one of both artistic beauty and geological utility. And 66 million years later, our world is better in both those ways, thanks to their evolutionary success.

Spiny Skins: Echinoderms

Although they do not inhabit freshwater environments, echinoderms were abundant in the seas of the Mesozoic. Echinoderms today include the starfish, sea urchins, sand dollars, and sea cucumbers of today's oceans—they have yet to invade freshwater or terrestrial environments, so all forms are shallow to deep marine (really shallow, in fact, as tide pools are a favored home for many of them). And they lived in similar environments during the Mesozoic. Echinoderms evolved early in the Cambrian period and had already diversified into unusual groups like eocrinoids, carpoids, and edrioasteriods. They are characterized by a water vascular system, which includes the tube feet by which starfish, for example, can slowly move across rocks and sand. The eocrinoids were related to the stalked, bottom-dwelling filter feeders, the crinoids, which were so abundant in the Paleozoic seas and continue in deep-sea environments today. Edrioasteroids appear to have been ecologically urchin-like forms. By the time of the dinosaurs though, most modern groups had appeared, as the Cambrian and Paleozoic evolutionary faunas had given up ground to the emerging modern evolutionary fauna in the seas—for the echinoderms, a fauna dominated by echinoids (sea urchins and sand dollars) and starfish. Such familiar modern forms occur in Mesozoic deposits around the world.

North of Como Bluff, Wyoming, where the "Bone Wars" (over dinosaurs) took place in the 1800s, the rolling plains of the Sheep Creek area expose rocks of the Middle–Late Jurassic–age Sundance Formation, which represents deposition of the Sundance Sea before the heyday of giant dinosaurs in western North America. These shallow marine rocks are one place where you can find limestones with Jurassic crinoid fossils; although crinoids were common in the Paleozoic and survive today, their fossils in the Mesozoic are rather rare. Another Jurassic marine deposit that has produced these is the La Voulte site in France, and they are known from some Cretaceous deposits as well (e.g., in the Niobrara Formation of Kansas and the Mancos Shale of Colorado and Utah).

Crinoids are animals, but from a distance, they are reminiscent of a flower or papyrus plant. The body consists of a calyx, which is an oval structure containing most of the organs and mouth. Several arms extend up from this (like petals of a flower), each with tube feet of the water vascular system extending out to capture food particles from the water currents. The arms and tube feet work the food down to the mouth. The calyx is attached by the down side opposite the mouth to a stalk composed of many circular plates with a hole in the middle—in stacked sections, the stalk plates look like rolls of Lifesavers or, because the outside edges of the individual plates are interlocking teeth, like stacks of thick poker chips. Isolated and stacked stalk plates and sections of them are common fossils in Paleozoic limestones and frequently indicate crinoids in the Sundance Formation. The Niobrara Formation and Mancos Shale form is *Uintacrinus*, a crinoid that has no stalk, only a calyx and arms (fig. 2.7A), and it was likely free-floating in large numbers.

Globally, sea urchins occur in many shallow marine deposits throughout the Mesozoic. These forms include spiny taxa similar to what you might see in a tide pool today (fig. 2.7*F* and 2.7G) and smooth but inflated rather than thin taxa similar to the sand dollars you might see on a sandy beach today (fig. 2.7*B* and 2.7*C*–2.7*E*). Starfish fossils are also known from all three periods of the Mesozoic (fig. 2.8A–2.8*E*), and we even have fossils of the traces they left on sandy sea bottoms (fig. 2.8*F*). The forms of these echinoderms of the Age of Dinosaurs are little different from modern starfish and sea urchins, and it is probably safe to assume that their ecologies were, in most cases, rather equivalent too. Clearly, the shores of Mesozoic seas, both rocky and sandy, were as colorful, diverse, and active as are our visits to the beach today. One can almost smell the breeze.

In Germany, Upper Jurassic rocks contain all the main modern groups of echinoderms: crinoids, starfish, brittle stars, sea urchins, and sea cucumbers.

2.7 Some echinoderms of the Mesozoic. (*A*) Complete specimens of the free-floating Late Cretaceous crinoid *Uintacrinus* (DMNH 6000) from the Mancos Shale of western Colorado. (*B*) Close-up of sections of feeding arms of *Uintacrinus* (FHPR 17811) from the same area. (*C*) Echinoid from the Lower Cretaceous of Texas (FHPR 2961). (*D–H*) Echinoids (sea urchins) of the Lower Cretaceous Edwards Limestone of Texas (FHPR 3079). Scale bars = 1 cm in *B*, 5 cm in *C*, 15 cm for *D–H*. *All photos by author.*

2.8 Some Mesozoic starfish and their "tube-feet prints" in the sands of time. (*A*) Middle Triassic *Trichasteropsis* from Germany. (*B*) *Eokainaster* from the Middle–Upper Jurassic Stump Formation of northeastern Utah. (*C, D*) Top (aboral) and bottom ("mouth") views of *Betelgeusia* (now *Coulonia*) from the Lower Cretaceous of Morocco (MHNT. PAL.2010.2.1). (*E*) Late Cretaceous *Metopaster*. (*F*) Starfish resting trace *Asteriacites* from the Late Cretaceous–age Mesaverde Group of Utah. Scale in *A* in centimeters, scale in *B* and *E* = 1 cm, scale in *F* = 5 cm. *Images in A and C–E courtesy of Daniel Blake; images in B and F by author.*

The success of echinoderms that we see in today's oceans and tide pools started early in the Mesozoic, probably around the time the dinosaurs were getting their start, during the slow recovery from the Permian–Triassic extinction after the early millions of years of the Triassic. They seem to have regained success fairly quickly and, in their more modern forms, shared the Mesozoic with most of the other main modern marine animal groups for the rest of the era.

Wing Gills: Hemichordates

Hemichordates are in the Deuterostoma with the echinoderms and the vertebrates we will see in this book, the three being more closely related to each other (especially by shared developmental characteristics) than they are to other invertebrate groups we meet in this chapter. Hemichordates include acorn worms and pterobranchs, the latter of which lived in tubelike colonies. The almost microscopic animals lived in the ends of the tubes and filtered organic particles from the seawater. During the Paleozoic, fossils of these animals called graptolites can appear in dark shales and look like miniature hacksaw blades, a tiny animal living in the upside cup of each of the "teeth" of the blade. These are very rare as Mesozoic fossils, but there are forms known from at least the Middle and Late Jurassic of France and Germany, respectively.

The Oldest Animals: Sponges

Other invertebrate groups appeared in the Cambrian or Precambrian (the latter prior to 542 million years ago) and are still in our oceans today. Metaphorically speaking, they sailed through the Mesozoic too. These include the sponges and cnidarians (see below). Sponges, of course, are the vase- to brain-shaped colonies of flagellated cells that work together to filter food particles from seawater. They are the simplest type of animal known and probably evolved a bit before the Cambrian during the Ediacaran or Cryogenian periods. There are about 5,000 species known today, but there appear to have been more species during much of the second half of the Mesozoic, from the Late Jurassic through the Cretaceous. Sponges are and have been important members of reef communities since the Age of Dinosaurs and were particularly common in the reefs of the Jurassic and Cretaceous around the world. The family Spongillidae, however, is unique in consisting of species that live in freshwater lakes and rivers, and it appeared during the Jurassic period. In fact, the new spongillid *Eospongilla* was described in 1999 from lake deposits in the Morrison Formation of Colorado, a tiny sponge that lived in the shallow lakes, potentially in the wading footsteps of dinosaurs such as *Apatosaurus* and *Stegosaurus*. Even sponges, of all animals, lived alongside the dinosaurs.

Stingers: Cnidarians

The cnidarians include jellyfish, sea anemones, hydrozoans, and corals. There are about 11,000 species today, and the group originated sometime around the Precambrian–Cambrian boundary 540 million years ago. The scyphozoans are the true jellyfish, of which there are three to four species known just from the Upper Jurassic shallow marine deposits of Germany. Colonial, sessile hydrozoans with a calcareous or other skeleton appear in the Late Cretaceous and are still around. Feathery colonial forms such as *Aglaophenia*, with each animal living in a tiny cup on the branching structure, can sometimes be found among poststorm beach debris in places like California and elsewhere. Although these animals are cnidarians and in a separate phylum, their colonial structure and mode of life are similar to some ancient hemichordates. Within Anthozoa, the Octocorallia (soft corals) include the sea fans of the order Alcyonacea. The sea fans, like the often-purple Venus fan *Gorgonia flabellum*, are common on coral reefs today and may have appeared in the Late Jurassic of Europe.

The scleractinian corals appeared around the Middle Triassic, well after the disappearance of the rugose and tabulate corals at the end of the Permian. Those latter groups had dominated the Paleozoic, but after their extinction, Earth was devoid of corals with hard calcareous skeleton structures for the polyp animals, and they may have only survived as sea anemone–like individuals. By the middle of the Triassic, however, corals seem to have developed skeletonizing ability again, and this is when the scleractinians appear. They diversified throughout the Mesozoic, particularly during the Late Jurassic when many reefs were developing, and they are the main builders of our gorgeous reefs today. These modern corals prefer depths of 50 m (164 ft) or less and water above 15°C (59°F), although they do best in depths around 20 m (65 ft) and temperatures of 25°–29°C (77°–84°F). They also do best in relatively clear water with plenty of sunlight. (Clear water is even more of a requirement for the sponges on the reefs.) Intriguingly, it appears that modern corals are adversely affected by turbid water less as a result of suspended sediment and more by the inorganic nutrients. And coral species that existed before the Early Cretaceous appear to have been more susceptible to turbidity issues, and tolerance has increased somewhat since then. These modern preferences can give some idea of the likely conditions in settings where we find fossil scleractinians in the Mesozoic.

Fossil corals of the order Scleractinia have been found around the world, on every continent, including Antarctica, a fact illustrating that continent's warmer ancient climate history. And their morphologies are similar to those familiar to scuba divers and snorkelers. In Jurassic and Cretaceous formations, there are fossil forms that are hemispherical (including with "brain coral" morphologies, known as meandroid), foliaceous, conical to mushroom-shaped, and digitate in structure.

As we will see, some of the fossil fish seen in coral reef–associated deposits of the Mesozoic also indicate that the reefs formed by these corals had some fish of rather familiar, reef-type shape swimming around them, even if they were among now-extinct groups. So, would snorkeling coral reefs of the Mesozoic have presented us with scenes broadly similar to those we encounter now, with schools of tang-like fish swimming among coral heads? To a large degree, yes. But, as we will also see, there were a few strange animals lurking out there that we'd always have a chance of encountering.

Lamp Shells: Brachiopods

Brachiopoda is a phylum of animals that appeared in the Cambrian and survives today, sometimes known as the "lamp shells." They are generally similar to bivalve mollusks in that they consist of an animal living mostly inside a pair of shells, but the plane of symmetry of the shells is opposite (between the shells in bivalves and through the center line of the shells in brachiopods). The animals are of different phyla too. Brachiopods have a soft-part coiled structure inside the shell known as a lophophore, with which the animals filter food out of the water they pump through the shells. The soft parts of bivalves are completely different. In contrast to bivalves, brachiopods are quite rare in the Mesozoic and all are marine.

Brachiopods were particularly abundant and diverse in the Paleozoic, although they have been far less common since then. In contrast to approximately 12,000 fossil species, there are fewer than 400 species today. Brachiopods reached the peak of diversity of about 900 genera in the Devonian period of the Paleozoic and then mostly declined from there, especially at the Permian–Triassic boundary. There are several dozen genera of fossil brachiopods reported from the Early Triassic from various continents, and through the Early to Middle Jurassic, their genus numbers rose from about 50 to more than 100; Jurassic forms are known from Africa, Europe, South America, Asia, and North America. There are also a handful of brachiopod genera from most continents (including Antarctica) known in Cretaceous deposits. Their diversity has hovered around 100 genera, give or take, since the Cretaceous.

Moss Animals: Bryozoans

The bryozoans are a mostly marine invertebrate group also known as "moss animals." These mostly tiny animals live in a small "cup" that they secrete, sometimes solitary but in others in large colonies. All have a lophophore (a structure they share with brachiopods) with an array of ciliated tentacles around the mouth, and the animals filter food particles out of the water as they remain stationary in their cups.

Bryozoans appeared in oceans late in the Cambrian and have survived up to today; their heyday occurred during the late Paleozoic when they were a common element of early reefs. You can find their fossils, which often look like tiny cargo nets or miniature corals in the rock, in limestones all over the Midwest and even in the deserts around Moab, Utah, but they were global in their distribution. They are also known as fossils in Europe, Russia, South America, and New Zealand in Triassic rocks; from Europe and Texas in the Jurassic; and from all over Europe, from North and South America, and from India, Tunisia, Egypt, South Africa, and Madagascar in the Cretaceous.

The four biggest bryozoan groups of the Paleozoic disappeared at the end of the Permian or Triassic, but one, the Tubuliporata (surviving today) made it through those episodes, diversified a little in the Jurassic, and then exploded in the Cretaceous. Tubuliporata took a large hit at the end of the Cretaceous and has only recovered diversity somewhat since then.

Within modern Bryozoa, the class Phylactolaemata has pioneered life in freshwater environments, and forms like *Cristatella* sport tentacles arranged in a U shape around the mouth. These animals sound a bit alien to our ears: they also possess colonial mobility in that their individuals (zooids, up to a couple dozen or more of them) live atop a common gelatinous mass with a mollusk-like foot on which the entire colony can crawl around. And, living in freshwater environments sometimes subject to drying up or freezing, the phylactolaematans have evolved the ability to go dormant in the winter and form hardened reproductive bodies, known as statoblasts, that can wait out the unfavorable seasonal situations and produce new zooids when conditions improve. Fossil phylactolaematans have been reported from an apparent encrusting colony from the Cretaceous of the Czech Republic and purported statoblasts of Triassic and Cretaceous age. Most of these reports have met with some skepticism, however, and it is possible that the first confirmed phylactolaematan fossils do not appear until the Quaternary. So, whether or not phylactolaematans inhabited ponds and lakes of dinosaur times remains a bit of a mystery.

The Geological Champions

The Mesozoic record of invertebrate fossils reveals to us a world surprisingly similar to our own in the form and composition of the species and ecosystems of these groups during the time of the dinosaurs. Obviously, it has been an "insect world" since those early days of the Mesozoic and continues today. What annoyed the dinosaurs and their tetrapod contemporaries were many of the same swarming insects we swat so often all summer in our backyards. The biostratigraphy of the era and the geological utility that goes along with that ability to correlate are largely a result of the mostly unsung invertebrate heroes of geology. The sudden extinction of ammonoids at the end of the Cretaceous and a few other groups along the way have helped us see the effects of environmental disruption, whether it is sudden and induced by rare events such as large extraterrestrial impacts, or slower and more mundane and due to other causes. Studies of diversification events and the rise of modern marine faunas have been largely based on these invertebrates also. Because invertebrate groups often have specific environmental parameters that they prefer, finding different types of animals in deposits with dinosaurs often informs us about the landscape settings that the larger animals occupied.

Whatever we know about dinosaur ecosystems and the environmental settings occupied by the animal groups that follow in the coming chapters, a significant amount of the information either originates from invertebrates or was enhanced by what we've learned from them. Although I admit that this book is guilty of underrepresenting the invertebrates to some degree—followed as this chapter is by a slew of vertebrates—the insects, chelicerates, crustaceans, mollusks, and other invertebrates of the Mesozoic are worth diving into more deeply, and there are sources that can do that more justice than we can here.

Notes

1. Stem taxa are those within a larger related group but outside (previously branched from) the more exclusive group defined by living representatives (which is the crown group). The crown group does not include *only* living taxa, however. A fossil taxon may be either part of the crown group or a stem taxon; it all depends on if it is most closely related to the common ancestors of the modern taxa or if it appears to have split off, within the broader group, before the ancestor of the modern representatives. *Anomalocaris* and its relatives, for example, are stem taxa because they are arthropods but are earlier-branched and more basal on the tree than the chelicerates and crustaceans, which survive today.

2. I can't help but think dinosaurs must have had as little use for ticks as we do.

3. This is my own interpretation of dinosaur traces described by Martin Lockley and others; my take is based largely on too many hours watching our backyard chickens dust bathing.

4. The blood of horseshoe crabs is used in the medical industry to check the purity of vaccines; the blood cells can identify possible contamination of the vaccine fluid by unwanted bacteria.

5. Of course, without cows during the Mesozoic, there would be no butter in which to dip the shrimp, crab, and lobsters. Alas . . . I also admit to a bit of hypocrisy here in that it appalls me that anyone would boil something alive. I don't like that I like lobster, but it is an amazing once-a-decade treat.

Sweet Delta Dawn

Fish

3

LYING ON MY STOMACH, my face in the dirt,[1] in a backyard in Grand Junction, Colorado, I realized paleontology was not always as glamorous as it was cracked up to be. In reality, it very rarely is—either glamorous *or* thought to be, I suppose. But in this case, I was about to get an unexpected thrill. My crew and I had just spent most of a day dry-screening about 100 m^2 of the six inches of top-soil dirt behind Susan Webster's house in northwestern Grand Junction. We had been trying to zero in on a Cretaceous-age marine fish that had lived in the Western Interior Seaway about 80 million years ago, one that had become buried in the bottom muds of the seaway that became the Mancos Shale of western Colorado. Shoveling the dirt into buckets and pouring it into screen boxes often used in archaeological digs, we gradually worked our way through an area about 10 m × 10 m, tracking which grid produced which bone fragments of the fish, and we had hit a concentration in the eastern third of the square, on the uphill side. Lying there now, digging through the Mancos Shale itself rather than the overlying dirt, we were slowly picking our way through layers of sea-bottom mud turned to stone, hoping to find unweathered bones in place. And with one push of an awl and lifting of a piece of shale—there it was, a pair of dentaries of the fish, with teeth in place, exposed to the sun for the first time in 80 million years (fig. 3.1). It looked like it belonged to a dinosaur. The deep, stout left and right lower jaws were compacted onto each other, right over left, and facing south; the teeth were conical, long, and straight, almost like those of a large, slender-toothed crocodile. The teeth were up to three inches long. A dinosaur-like jaw but one from a fish. Definitely the world of the Cretaceous.

We knew we were on the trail of a fish when we arrived that April. Susan had brought a small paper cup of fossil fragments that her nephew had found almost ten years earlier to the Museum of Western Colorado and said they came from her backyard. Among those fragments was a vertebra very clearly from a fish, but it was an unusually large one. Not your standard record-setting trout—much larger. When we asked where she lived, it was a part of town prime for large dinosaur-age fish—she clearly lived in an area that had bedrock of Mancos Shale. Not the older rock close to Colorado National Monument, and not the thick river gravel deposits from close to town near the Colorado River. We determined to check the site out, which is how, along with significant luck, I was the one who first came face-to-face with the giant jaws now under my nose in Susan's backyard.

Table 3.1. Fish market Scrabble

Major Groups		Minor Groups and Characteristics		
Chondrichthyes (cartilaginous fish)	Sharks	Torpedo-shaped body, carnivorous with mostly very sharp teeth		
	Skates	Flattened body shape; lay eggs, no tail spines, retain dorsal fin		
	Rays	Flattened body shape; live birth, tail spines in most, reduced or absent dorsal fin (includes sawfish and guitarfish)		
Osteichthyes (bony fish)	**Sarcopterygians** (flesh-finned fish)	Lungfish	Lungs, tooth plates	
		Coelacanths	Secondary (three-lobed) tail, vestigial lungs	
	Actinopterygians (ray-finned fish)	Amioids	Bowfins (Amiidae)	
		Bichirs	Surviving archaic ray-fins with lungs and a long dorsal fin	
		Macrosemiids†	Archaic ray-fins with thick scales	
		Palaeoniscoids†	Archaic ray-fins with eyes placed relatively far forward on the skull, probably paraphyletic; includes coccolepids	
		Pycnodontids†	Archaic ray-fins with blunt, crushing to hook-shaped teeth and an often compressiform body shape	
		Redfieldiiforms†	Archaic ray-fins with thick scales and fusiform body shape	
		Semionotids†	Archaic ray-fins with short spines anterior to the dorsal fin	
		Sturgeons	Large scaleless fish (Acipenserids)	
		Teleosts	Ichthyodectids†	Large, carnivorous stem-teleosts
			Pholidophoriforms†	Fusiform to elongate stem-teleosts

Note. Many of the fish group names encountered in this chapter may be unfamiliar, even if the groups are not extinct. This is a guide to some of the terms used in this chapter. † = extinct groups

The fish that produced these jaws and the vertebra that had led us to this location is *Xiphactinus*, a large predatory fish of those ancient seas that is best known from deposits of the same age in Kansas (fig. 3.1). These more complete specimens indicate that the fish pushed 6 m (20 ft) in length! It was a member of an extinct group of mostly large predatory fish, though

***Facing,* 3.1** The giant ichthyodectid fish *Xiphactinus* from the Western Interior Seaway of the Late Cretaceous in North America. (*A*) The "fish in a fish" specimen from the Sternberg Museum at Fort Hays State University in Kansas. A nearly 5.8 m (19 ft) *Xiphactinus* with a recently ingested fish of nearly 1.8 m (6 ft) fish inside it. (*B*) *Xiphactinus* from the Niobrara Formation of Kansas (DMNH 1667). (*C*) The Webster Fish Quarry site in the Mancos Shale near Grand Junction, Colorado, which produced a *Xiphactinus* skull and vertebrae in 2012. (*D*) Right dentary of Webster *Xiphactinus* as it appeared when uncovered in the field (with consolidant added). Scale in centimeters. (*E*) Webster *Xiphactinus* (MWC 8064) previous display at the Museums of Western Colorado. Included here are left and right maxillae and premaxillae, left and right dentary, vertebrae, fin spines, and other cranial and postcranial bones. (*F*) Left and right dentaries of MWC 8064 after preparation. Lens cap approximately 5 cm (2 in) in diameter. (*G*) Premaxilla as found in the field. Lens cap approximately 5 cm (2 in) in diameter. (*H*) *Xiphactinus* feeding. Painting by Dan Varner and courtesy of Mike Everhart. *All photos by author, except A from the collection of the Sternberg Museum of Natural History, Hays, Kansas, and D by ReBecca Hunt-Foster.*

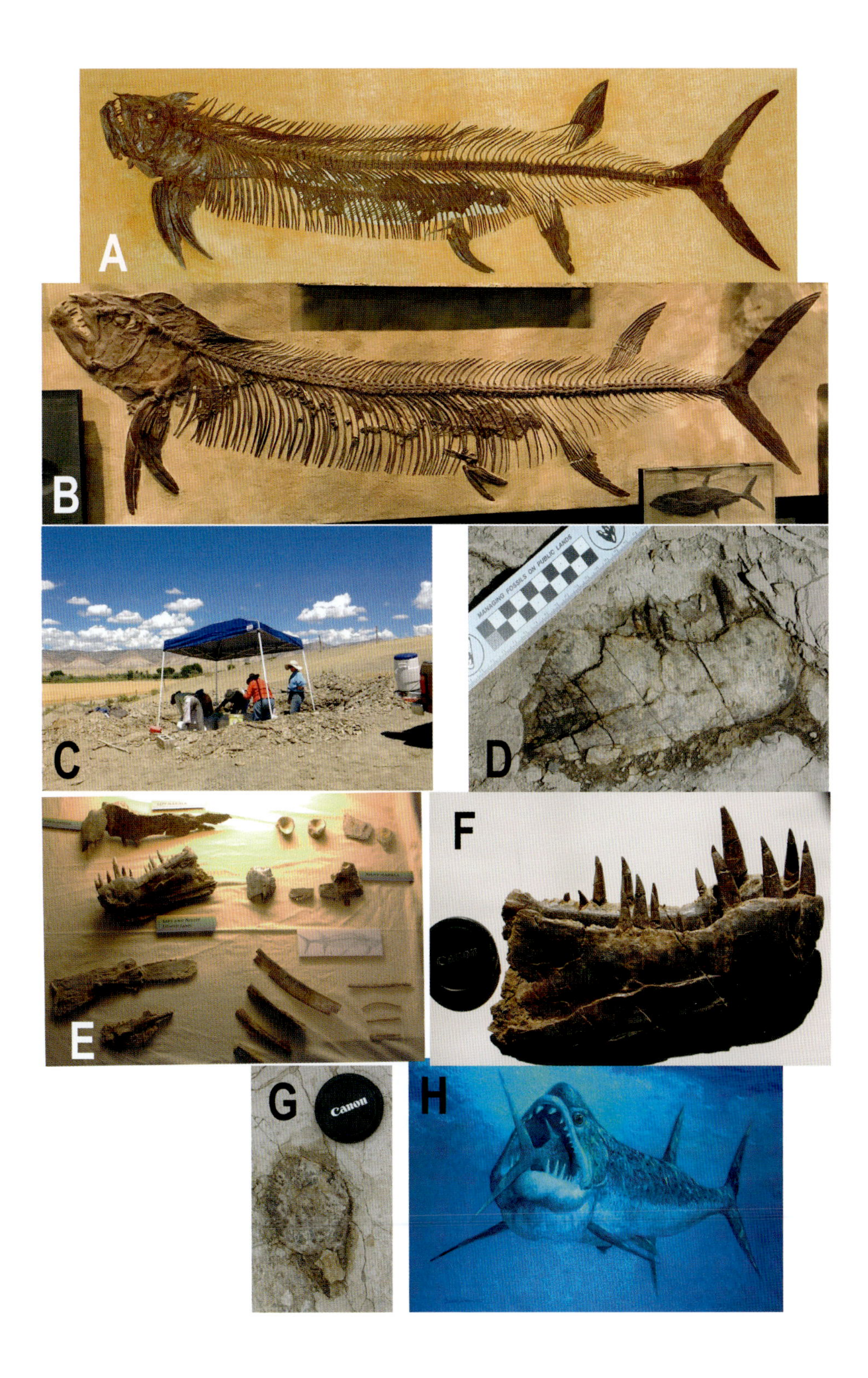
A
B
C
MANAGING FOSSILS ON PUBLIC LANDS
D
E
F
Canon
G
Canon
H

it was the largest. A famous specimen at the Sternberg Museum in Hays, Kansas, indicates that *Xiphactinus* sometimes ate fish, nearly whole, that were up to 1.8 m (6 ft) long (fig. 3.1A). This was a fish that was as dangerous as the sharks of the time and fit in quite well in the Age of Dinosaurs. It demonstrates that not all the wonders of the Cretaceous were of the giant, reptilian, and land-dwelling variety. Some truly colossal monsters of the time were in fact sea monsters related to our modern fish friends. Fish may be obvious companions of the dinosaurs in the Mesozoic world, but they do have their stories to tell us.

To the Campanian Fish Market

Years before that *Xiphactinus* work, I had my first significant encounter with the world of Cretaceous fish, at the end of that drive to Colorado with Dave Archibald. I eventually came out of my geological timescale trance in the International as we roared down the interstate, and I was a somewhat different student than I had been an hour earlier. I again began perceiving the cars, interstate, and desert around us. I was familiar with the territory, having come this same way many times over the previous year working on my undergraduate thesis project in Cambrian rocks in the Mojave. But the Cretaceous was going to be new to me at that point. I'd seen a few Cretaceous fossil localities in California, but they were all rich in marine invertebrates only. We were headed to mostly terrestrial and freshwater deposits in what had once been Laramidia, the western half of North America that had been separated from Appalachia (the eastern half) by the Western Interior Seaway, which at its maximum in the Late Cretaceous stretched across what are now the Rocky Mountain and Great Plains states and provinces, connecting the Gulf of Mexico to the Arctic Ocean in one continent-splitting and -flooding sea. Where we were headed, Dave explained somewhere along that drive, had at the time been a Mississippi-like delta of a river or rivers draining the east side of Laramidia into the western shoreline area of the Western Interior Seaway. We were focusing on the Williams Fork Formation of the Mesaverde Group, which was the deposited sediments of rivers, swamps, floodplains, and estuaries of that delta area along the ancient seaway shoreline. Lounging or prowling around that swampy delta had been duck-billed and ceratopsian dinosaurs, as well as albertosaurine tyrannosaurids. There were a lot of other animals, of course, as we would see in the coming weeks, but those were the big and flashy ones. Out in the seaway were sharks, giant fish, and marine reptiles such as mosasaurs and elasmosaurs. In the rocks we would be working in, we would probably not see any marine reptiles, but some of the sharks and fish frequented both freshwater and marine waters.

After a lightning- and rain-filled night camping at Capitol Reef National Park in Utah, we headed out on the last leg of our trip to the field area on a beautiful morning in early July. We arrived in Rangely, Colorado, midafternoon and did some grocery shopping before arriving at our camp spot among the juniper trees on a plateau of the Douglas Creek Arch, a geological upwarp that brings Cretaceous rocks up to the surface between

A Summer in the Field
Late Cretaceous Exploration

3.2 Scenes from the Upper Cretaceous of Colorado, hunting for dinosaurs and beast companions in 1989. (*A*) Outcrops of the sandstones and gray mudstones of the Mesaverde Group (Williams Fork Formation). (*B*) Sunset among the piñon and juniper. (*C*) The crew after a hike, *left* to *right*: Paul Majors, the author as a young pup, leader Dave Archibald (rehydrating), and Matt Colbert (caffeinating). Photo by Gloria Bader. (*D*) Block of sandstone with turtle and hadrosaur elements. (*E*) Sandstone slab with an imprint of a *Sabalites* palm frond, with rock hammer and antler for scale. (*F*) Turtle shell fragment, crocodile tooth, and tooth of a guitarfish. (Same field area but actually collected in 2012.) (*G*) Home for a month . . . among piñon and juniper. (*H*, *I*) Quarry shots of a hadrosaur (duck-billed) dinosaur (SDMNH 38229) during excavation. *All photos (except C as noted) by author.*

Cenozoic units exposed to the east and west in the Piceance and Uinta Basins. Camp was a nice home away from home, although it was my first official paleo field camp setup, so I didn't have much at the time to compare it to.[2] Parking spot in the sand, equipment tent, kitchen and dining shades enclosed with antibug screens and both connected by a screened vestibule, dining area complete with table and chairs, individual tents scattered among the junipers a few tens of yards out from the common area—a pretty typical arrangement and a very comfortable one as dry camping goes. It was our home for a little over a month, and during that time we welcomed to it five geologist and paleontologist colleagues, friends of Dave's from town, plus scorpions, rattlesnakes, and horned lizards (fig. 3.2). At least we were there late enough in the season to avoid the biting midges.

I was on this expedition thanks to blind-luck timing and people I happened to know who happened to know someone else. I had waited until the end of my second year in college to decide whether to major in marine biology, history, or geology, finally settling on the study of the Earth. It wasn't until nearly the end of my senior year that I finally decided that indeed among all the specializations within geology, the one that most motivated me was the least practical—paleontology.

On various field-mapping exercises during my geology major, we would be in the Mojave Desert mapping Paleozoic formations overturned by imbricated thrust faults, and whenever our professor caught up to my mapping partner and me, she would find Jim measuring strikes and dips and drafting his field map, while I was pounding away splitting limestone or shale in search of just one more horn coral or trilobite. I enjoyed the mapping and did just fine at it, but a chance discovery of a fossil while out "geologizing" (to steal a term from John Wesley Powell—or was it one of his impatient crew members?) led me down a whole rabbit hole of obsessive pursuit of the ever-elusive "perfect one." She'd almost have to drag me back to my assignment. This was the indication of what I should probably focus on if I wanted to pursue a specialty in graduate school. Despite that, I got into graduate school in a program that would have had me studying marine shale units of the Cretaceous in Utah and surrounding areas, units that produce plenty of clams, oysters, and ammonites, but those were simply not part of the project for which I would have been funded.

At nearly the end of my last term in college, I balked. Worrying that if I went through with the plan I had at that time and ended up always wondering if I could have pursued my true, if impractical, passion, I bailed out on my graduate school arrangement and decided to apply to a paleontology program instead. Better to try rather than spend a career wondering, I figured. It would be easy enough to go back to geology if the paleontology-as-job angle didn't pan out, but to attempt to break into paleo after training as a stratigrapher or oil geologist would be a little tougher.[3]

So I went to talk to Don Prothero, my undergraduate adviser at the time, and gave him the bad news that I was going to apply to a vertebrate paleontology program, the least practical specialization within an impractical specialization. He told me I was crazy and gave me one of the two "scare the students" lectures he prides himself on, this one a litany of reasons why no one in their right mind goes into vertebrate paleontology. Not an angry tirade, of course. A gentle heads-up, a wake-up call to dispel any misconceptions. The reality of almost no jobs, bad pay, dead-end job tracks when you do find one, you usually do something else for your actual work and your paleo research is merely tolerated by your institution, on and on. I said I didn't care. At that point, he said, "Okay, here's what you need to do . . ." Don is a vertebrate paleontologist. He just wants to make sure none of his students is "misled" into his field. He was very conscientious about not letting any of us go into paleontology with any illusions about what we were setting ourselves up for.

I was accepted into that graduate program and started it about a year and a half later, taking one academic year off as a break, during which I worked for a geological engineering company, mostly doing postearthquake mitigation studies. But one thing Don did for me as I was nearing graduation was to get me started on a project just for the experience. In our geology department, we had a few buckets of Bug Creek Anthills matrix ready to have tiny fossils picked out of it—mammals, fish, lizards, and dinosaur teeth from the very end of the Age of Dinosaurs and the beginning of the Age of Mammals.[4] Don was in his late thirties at the time and had done his graduate work at Columbia University, specializing in fossil rhinos. He had taught at several colleges, so he knew a few people in the profession, which is a rather small community to begin with. Even among today's larger community, most vertebrate paleontologists know each other or are only one degree of separation from pretty much everyone else, even globally. Early in my senior year when I was looking for a thesis project, Don called Dave in San Diego. Dave broke the news to me through Don that I'd probably have a tough time sorting and identifying the fossils with no previous training in Bug Creek material or others of that particular age, so I needed to find another project. That's when the Cambrian beckoned.

But late in my senior year, when I was nearly done with my Cambrian thesis project, Don gave Dave another call, knowing I could use paleo-specific field experience before grad school if I could get it. As it happened, Dave was headed out to the Cretaceous of western Colorado that coming summer and had one more crew spot open. I jumped at it. Don must have assured Dave that I wasn't completely inept in the field because Dave didn't know me at all and was taking a bit of a chance bringing me along. I didn't know it yet, but it would be hard *not* to find fish where we were going.

By the time we headed out for our first field day that July, I still had no idea what to expect but was looking forward to seeing what we found. What we found first were Barrier Canyon–style pictographs, the "Carrot Men," archaic Native American rock art illustrations from up to 4000 years ago. Then we went to a site that had been found about three years earlier and at which we split open sandstone to find crocodile teeth, turtle bones and shell pieces, a mammal tooth, and the teeth and scales of fish (fig. 3.2). This serendipitous first dip into paleontological fieldwork was a crash course in microvertebrates and fish in particular. I first learned the process of wrapping dinosaurs in plaster and burlap and the art of screenwashing on Dave's Cretaceous trip. My first month living in the field must have had some influence on future field project logistics for me as well.

In this fieldwork in the Cretaceous rocks of Colorado, fish fossils were abundant. Bowfin teeth. Bowfin vertebrae. Gar scales, teeth, and skull pieces. Bowfins and gars occur in the large, slow rivers of the Mississippi and its tributaries (along with other eastern river systems and the Great Lakes) today. And then the Williams Fork Formation also contains rare elements of the fish fauna like lungfish (just found recently) and pycnodontids. Sharks, rays, and guitarfish were diverse but not tremendously abundant as fossils except at certain sites. We saw fish scales and teeth in anthills, the ants having built their colonies and lined their hills with tiny pebbles the size of the small scales and teeth of the ancient fish. Gathering these from the surrounding area and transporting them home, the ants concentrated the fossils. We found the same fossils and larger vertebrae in the sandstones of the area, ancient concentrations of bone debris in the bottoms of river channels now turned to semiarid, high-desert stone. We often found fish fossils lying on the surface of mudstone slopes as the first indicators (along with turtle shell) of sites in the carbonaceous deposits of old Cretaceous floodplain ponds. Whether we were quarrying hard sandstones with chisels, picking fossils off the surface while crawling on our stomachs, or screenwashing matrix in unused cattle tanks, we found fish fossils.

Fish were pretty much everywhere, and whenever we returned to the same area in the years since, the trend has continued. We are still finding fish daily, and this is not unique to freshwater Upper Cretaceous rocks in general. Most of the terrestrial rock units, the formations of the Rocky Mountain region today—of what was the east coast of Laramidia, or the western shoreline of the Western Interior Seaway during the Cretaceous—contain many of the same fossils. Units like the Aguja, Kirtland, and Kaiparowits formations to the south of where we were in Colorado (Texas, New Mexico, southern Utah) and the Judith River and Horseshoe Canyon formations to the north (Montana, Alberta) contain sharks, bowfins, gars, and/or sturgeons and other fish in fairly significant numbers. Just after the time of the Williams Fork Formation, in the latest Cretaceous, rocks known as the Hell Creek and Lance Formations were laid down well above the now-buried Mesaverde Group—and they also contained the fossils of abundant bowfins and gars, some of the same genera, in fact. There were also sturgeons and, in South America, ancient catfish. And it wasn't just bony fish either. These Cretaceous units almost all contain significant numbers and diversities of cartilaginous fish as well, such as sharks, skates and rays, and guitarfish. The guitarfish *Myledaphus*, for example, is widespread in the Late Cretaceous of North America and Asia, mainly from teeth, although more complete specimens are known (fig. 3.3).

Out in the marine waters of the Western Interior Seaway at the time were more sharks and a wide variety of fish, including some quite large ones. As

3.3 Ventral view of the extremely rare, fossilized cartilaginous skeleton of the freshwater guitarfish *Myledaphus* (TMP 1998.062.001), found in the Late Cretaceous–age Dinosaur Park Formation in Alberta. (*A*) Approving Utah state paleontologist Jim Kirkland for scale. On the *Myledaphus* skeleton, note snout on right, along with pectoral and pelvic fin bases in center. (*B*) Close-up of tooth battery of interlocking, flat teeth for crushing clams and other hard prey. These teeth are common as isolated fossils in Upper Cretaceous continental deposits of North America and also occur in Asia. *Photos courtesy of Jim Kirkland and taken at the Royal Tyrrell Museum of Palaeontology in Drumheller, Alberta.*

mentioned, *Xiphactinus* grew to nearly 6 m (20 ft). *Enchodus* was a small to medium fish, depending on which of the dozen or so named species we are talking about. Although the larger ones were a bit smaller than *Xiphactinus*, they were still large predatory fish with sharp conical teeth, especially in the anterior fangs in the fronts of the upper and lower jaws. *Enchodus* has been reported from South America, Asia, Africa, and Europe, and in North America, it is known from numerous states and provinces in the East, Southeast, Southwest, and Rocky Mountain regions of the United States, Mexico, and Canada. Particularly good specimens of the large (1.5 m [5 ft]) species *Enchodus petrosus* come from the Niobrara and Pierre Formations of the Great Plains, whereas jaws of significantly smaller species are known from the Mancos Shale near Fruita and Moab in Colorado and Utah, among other formations and areas.

Ancient Fins Up!

Fish are some of the most common fossils found in almost any time period. They originated early in the Paleozoic, about 270 million years before the dinosaurs. Starting out as boneless and jawless chordates that nonetheless possessed cartilaginous vertebrae, gills, dorsal nerve cords, and myomeres, they first appeared in the Cambrian period alongside trilobites and the bizarre arthropod predators such as *Anomalocaris* and *Yohoia* in forms such as *Haikouichthyes* and *Metaspriggina*. By the Ordovician, fish such as *Arandaspis* and *Sacabambaspis* had reached slightly larger sizes and were still jawless and cartilaginous internally, but many had developed external armor coverings of many tiny bony plates. Eventually, the jaws and internal skeletons began ossifying into bone rather than cartilage and they began developing hard teeth. The bony plates of early jawless fish and heavily armored "placoderm" fish of the Silurian and Devonian mostly disappeared in favor of scales. The bony fish, or Osteichthyes, were born in the Silurian with fishes such as *Lophosteus*, *Psarolepis*, and *Andreolepis*.[5] Around this time, this group split into the ray-finned fish (or Actinopterygii), most of the familiar fish today, and the lobe-finned fish (Sarcopterygii), which includes the modern and ancient lungfish and coelacanths, for example. At some point early on, the cartilaginous sharks and rays split off as well, with the spiny-finned acanthodian fish, such as *Acanthodes*, scattered among the earliest, basal Osteichthyes and Chondrichthyes. All of this happened fairly early in the Paleozoic, so most of these groups were diversifying greatly in the second half of that era. By the time dinosaurs finally arrived, many of the fish belonged to modern groups, although some older forms were still around. But it could be argued that the freshwater and marine realms before, during, and after the time of dinosaurs have always been in an "Age of Fish." Osteichthyes suffered a moderate extinction at the Permian–Triassic boundary (measuring at the genus level) but then appears to have diversified significantly in the Triassic.

Fish were here when dinosaurs evolved, and they continued after the nonavian dinosaurs disappeared. Among the first dinosaur companions in the rivers and lakes around which early dinosaurs lived were sharks (along with skates and rays), coelacanths, lungfish, and archaic ray-finned fish (e.g., redfieldiiforms, palaeoniscoids, pycnodontids, and semionotids). Coelacanths, of course, are the lobe-finned fish thought extinct since the Cretaceous that reappeared in a fishing boat off Madagascar in the 1930s, and lungfish are the lobe fins known from freshwater environments of South America, Africa, and Australia. The remaining fish of the early dinosaur era listed above are primitive ray-finned fish (actinopterygians). A number of these are known from the Chinle Formation of southern Utah in places like Lisbon Valley where Andrew Milner and Sarah Gibson have been working on some remarkably complete freshwater semionotid, redfieldiiform, palaeoniscoid, and coelacanth fish from the time of the first North American dinosaurs. The spine-headed shark *Xenacanthus* and spiny-finned hybodont sharks can be relatively abundant in microvertebrate sites in the Chinle Formation, and lungfish show up occasionally too, even tiny ones, as I saw firsthand that July day in southern Utah a few years ago.

During the Jurassic, there were still palaeoniscoids in dinosaur deposits in lakes and rivers, but there were a few teleosts (more advanced ray-finned fish) too. Southern continents have produced numerous Jurassic palaeoniscoids, semionotids, pycnodontids, amioids, and teleosts, many of which have been studied by Adriana López-Arbarello and others. Some of these fish are well preserved and nearly complete.

In isolated Rabbit Valley, west of Fruita, Colorado, quiet Exit 2 off Interstate 70 dumps you onto dirt roads to nowhere—nowhere if you're interested in "places." Here, the dirt roads only lead to other dirt roads—Utah in one direction, the Book Cliffs in another, and, in half a dozen miles or so, the Colorado River, traveling its "lonely and majestic way" through Ruby Canyon, in another. The dirt roads go for dozens to hundreds of miles. But you need only go a quarter mile or so to get to the Mygatt-Moore Quarry, the site that produced the crayfish we discussed in chapter 2. The same beds that produced the crayfish also produce the swimming, vertebrate true fish at this site, some of the few articulated fish in the Morrison Formation. Among these fish are the coccolepid palaeoniscoid *Morrolepis schaefferi*, the possible semionotid "*Hulettia*" *hawesi*, and a more derived teleost similar to *Leptolepis* (fig. 3.4A–3.4C). Most modern fish are teleosts, so *Leptolepis* was one of the early representatives of that group. Coccolepids and most of the other groups known from the Morrison are also found in deposits of similar age in other parts of the world. Elsewhere in the Morrison Formation, near Cañon City, Colorado, quarries at Temple Canyon have produced a coccolepid palaeoniscoid, a possible semionotid, an amioid (a relative of the modern bowfin), a pholidophoriform teleost, and other actinopterygians (fig. 3.4*D*). Temple Canyon also produced the most complete lungfish skull in the Jurassic of North America. Both of these Morrison fish faunas were recovered from deposits representing small lakes. Elsewhere in the Morrison Formation, as I found while a student at the University of Colorado, fish are found in pond and river deposits over much of the

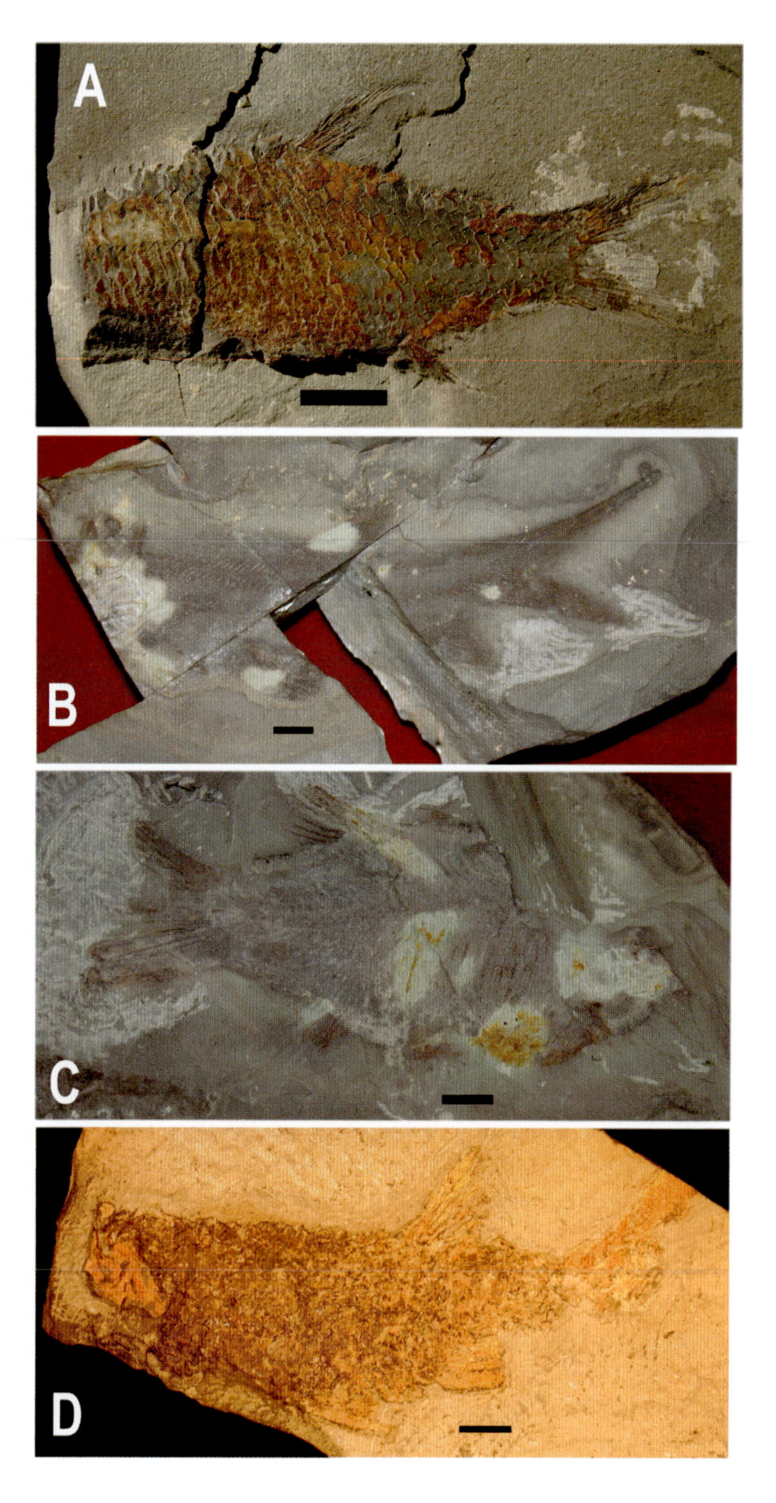

3.4 Freshwater fish of the Upper Jurassic Morrison Formation in North America. (*A–C*) Fish of Rabbit Valley in western Colorado. (*A*) The possible semionotid "*Hulettia*" *hawesi* (MWC 5564). (*B*) The coccolepidid palaeoniscoid *Morrolepis* (MWC 440). (*C*) The teleost cf. *Leptolepis* (MWC 3722). (*D*) An indeterminate actinopterygian from Temple Canyon, central Colorado (DMNH 58009). All scale bars = 1 cm. *All photos by author. Specimens in images A–C located at Museums of Western Colorado.*

3.5 Marine fish I. Amiid and ophiopsid fish from the Late Jurassic of Germany (Solnhofen Formation). (*A*, *B*) Amiids. (*A*) *Solnhofenamia*, CM 4731, total length ~ 41.4 cm (16.6 in). (*B*) *Amiopsis*, JME-ETT 284, scale bar in centimeters, from Ettling. (*C*, *D*) Ophiopsids. (*C*) *Ophiopsiella*, JME-ETT 1896, scale in centimeters, from Ettling. (*D*) *Ophiopsis*, JME-ETT 4109, total length ~11 cm (4.4 in), from Ettling. *Image in A from Grande and Bemis (1998), courtesy of Lance Grande; fossils in B–D property of the Bishops Seminar St. Willibald Eichstätt. Access by the Jura Museum Eichstätt, photos by Martin Eber.*

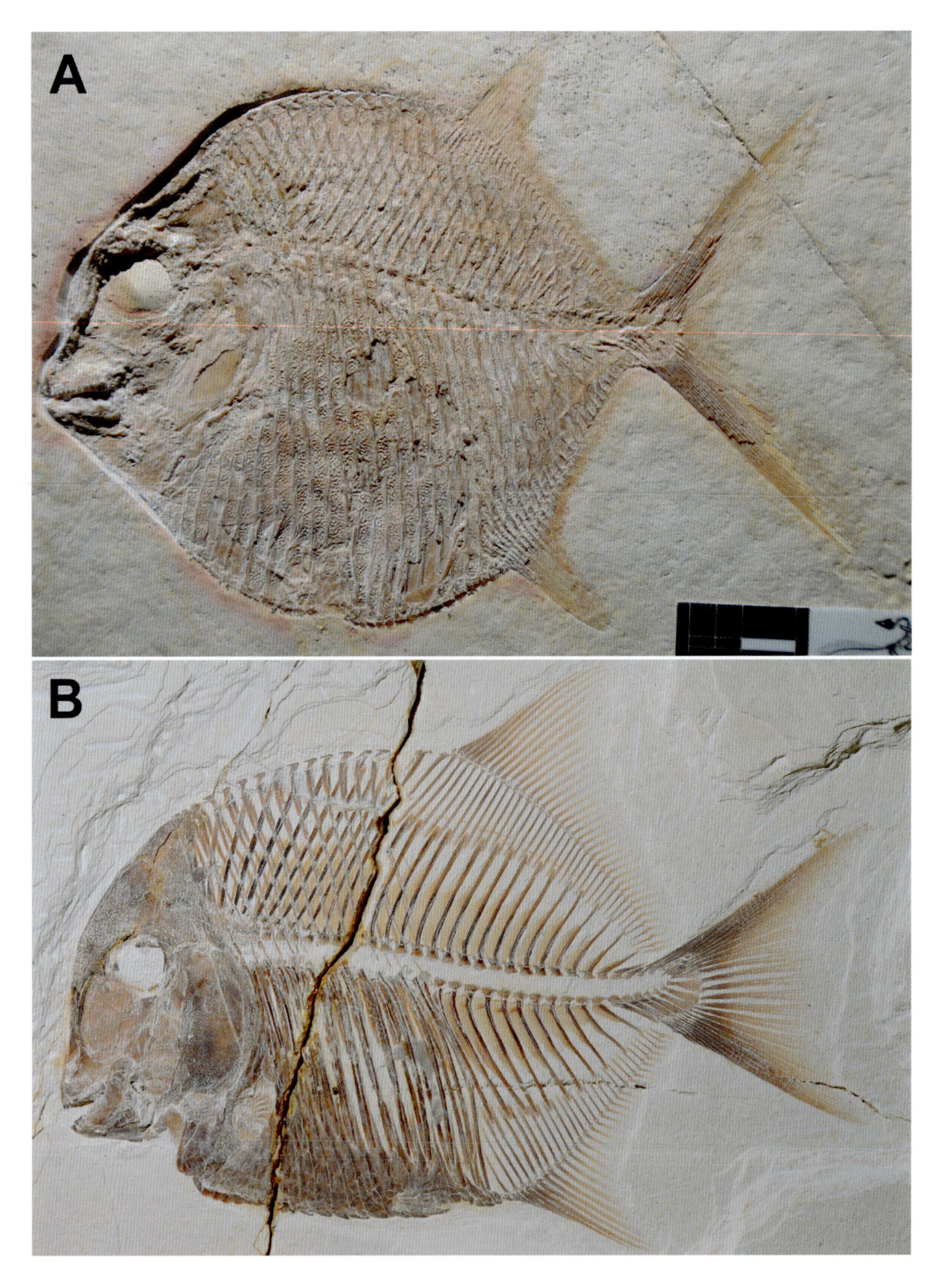

3.6 Marine fish II. Pycnodontiform fish from the Late Jurassic of Germany (Solnhofen Formation). (*A*) *Gyrodus* from Eichstätt (JME-SOS 4303). (*B*) *Turboscinetes* from Ettling (JME-ETT 2749). Scale bar in *A* in centimeters. *Fossils property of the Bishops Seminar St. Willibald Eichstätt. Access by the Jura Museum Eichstätt, photos by Martin Eber.*

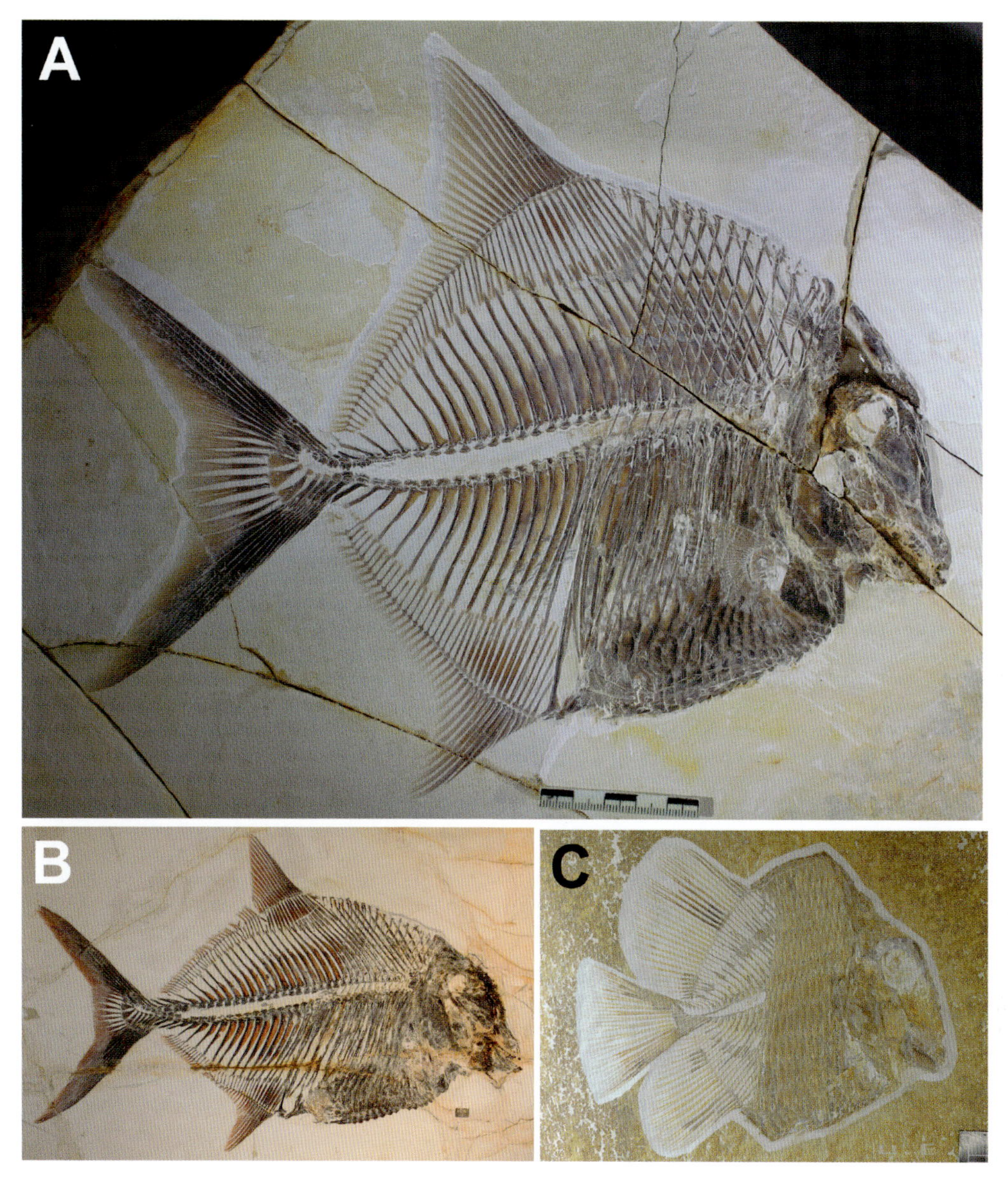

3.7 Marine fish III. More pycnodontiform fish from the Late Jurassic of Germany (Solnhofen Formation). (*A*) *Proscinetes elegans* from Ettling (JME-ETT 876). (*B*) *Proscinetes bernardi* from Ettling (JME-ETT 250). (*C*) *Macromesodon* from Solnhofen (SNSB-BSPG 1955 I 82). Scale bar in *A* in centimeters. Fossils in *A* and *B* property of the Bishops Seminar St. Willibald Eichstätt. *Access by the Jura Museum Eichstätt, photos by Martin Eber. C courtesy of the Bayerische Staatssammlung für Paläontologie and Geologie.*

3.8 Marine fish IV. Various fish from the Late Jurassic of Germany (Solnhofen Formation). (*A*) The ichthyodectiform *Thrissops* (JME-ETT 74). (*B*) The aspidorhynchid *Belonostomus* (JME-ETT 3912). (*C*) Detail of a different *Belonostomus* (JME-ETT 123a) with a prey fish still in its mouth. (*D*) The semionotiform *Macrosemimimus* (JME-ETT 978). (*E*) The salmoniform *Orthogonikleithrus* (JME-ETT 1945). (*F*) The macrosemiid *Notagogus* (JME-ETT 116). Scale bar in *F* = 1 cm. *All specimens from Ettling locality. Fossils property of the Bishops Seminar St. Willibald Eichstätt. Access by the Jura Museum Eichstätt, photos by Martin Eber.*

geographic extent of the unit, from scales and jaws of amioids in southern Utah to northern Wyoming to isolated round, flat teeth of pycnodontids in northern Utah.

In the 1800s, a rail worker named William Reed discovered the dinosaur beds at Como Bluff, Wyoming, and began working them for paleontology professor O. C. Marsh at Yale University. In the summer of 1879, Reed discovered what he referred to as Quarry 9, a site that has produced many mammals and other microvertebrates out of the Morrison Formation. Among the material that Reed and his crews collected from Quarry 9 that does not get a lot of attention but is preserved at the Smithsonian Institution's National Museum of Natural History is a box or two containing hundreds of vertebrae of amioid fish, indicating that bowfin-like taxa prowled the waters around which lizards, mammals, and small dinosaurs scampered. More evidence that the freshwater settings of the Late Jurassic were not tremendously different from those we know today.

Shallow marine deposits in the Upper Jurassic of Germany have produced coelacanths, sharks, rays, ratfish, amiids and ophiopsids (fig. 3.5), and pycnodontid reef fish (figs. 3.6 and 3.7), plus semionotids, macrosemiids, teleosts, and others (fig. 3.8). The preservation of these fish in the Solnhofen deposits at sites like Eichstätt and Ettling in Germany is spectacular. These fish lived in and around coral reefs scattered among islands on the flooded European continent bordering the western Tethys Sea. Some probably lived in deeper water. The fossil pycnodont *Gyrodus* from this deposit (fig. 3.6A) had an overall shape similar to modern reef fish such as the yellow tang (*Zebrasoma*). The teeth of *Gyrodus* are rather blunt, suggesting that it crushed hard food items, possibly along the reefs. And amiids, pycnodonts, semionotids, and teleosts also occur in the Morrison Formation and other deposits elsewhere. We see from these Late Jurassic fish faunas that some of the same groups of fish had representatives in both freshwater and marine settings of the time.

Back to the Cretaceous

By the Cretaceous, the more advanced teleosts had taken over the seas, lakes, and rivers in terms of abundance. In Brazil, the Early Cretaceous–age Romualdo Formation of the Santana Group has produced quite a fish fauna. This relatively thin geological unit has nevertheless yielded turtles, crocodiles, dinosaurs, and pterosaurs in addition to the fish. The fish include a bowfin (amioid), an ichthyodectid, a semionotid, a guitarfish, and 15 other species. The most unusual aspect of the fish out of the Romualdo is that they are often preserved in three dimensions—I don't mean the bones are in 3D, I mean the whole fish is there in all its scaly, torpedo-shaped glory, as if someone took a live fish, slapped it in the rock, and petrified it as is (as was?). This preservation resulted from early phosphatization and mineral diagenesis while bacterial action was still ongoing in the body cavity, resulting in the fish bodies forming the cores of calcareous concretions that retained lithified soft tissue and organ structures intact and in three dimensions for 100 million years. The exceptional preservation seems to have occurred after mass-mortality events, possibly resulting from severe salinity fluctuations in the shallow Romualdo sea, possibly near or in an estuary or other nearshore setting where such conditions are apt to vary. Diverse fish faunas are common in marginal marine to fresh-brackish water settings in the Cretaceous around the world.

Cairo's position at the upper end of the Nile Delta gives it (along with a bit of smog from the cars of the city's nearly 10 million residents, 20 million in the metro area), counterintuitively, one of the most humid atmospheres I've ever seen. Red-sunrise mornings of opaque to barely translucent particulate-enhanced fog often give way by midafternoon to clear skies except for the slight haze of automobile pollution. The water supply of the river feeds the roots of an abundance of plants and a "paradise on Earth" of greenery (to quote fictional excavator Sallah), even in the middle of the city. But the contrast between humid air, water, and the verdant valley of the Nile and the completely barren, sand and rock desert, nearly resembling Mars, is stark, and it often transitions from one extreme to the other in a space of literally yards, especially as one moves south along the river valley. Chugging upstream through Luxor to Aswan, one passes mile after mile of low-lying farmland and date palms lined on either side beyond the fields by blinding sand, limestone, and sandstone outcrops; this farmland has been cultivated for more than 200 generations. Water and life cling to the river while seemingly nothing can survive for long beyond that. The remnants of Egypt's spectacular ancient culture survive in that desert partly because of the desiccating environment in which temples, tombs, villages, and mummies are sometimes found, those in the river valley often having been consumed by subsidence into mud or decay at the hands of wet environments.

But in Egypt, the Nile and water are life. And in that river water are fish, some of them quite large. Forms such as the Nile perch, known in African lakes and rivers beyond the Nile, can grow up to nearly 2 m (6.5 ft) long and weigh up to 180 kg (400 lb). The catfish get up to 1.5 m (5 ft) long; the lungfish *Protopterus* grows to 2 m (6.5 ft) as well. Tigerfish, while not as large, have disturbingly large teeth that remind one a little of a *Xiphactinus*; thankfully, the Nile species is smaller than one in the Congo that gets up to 1.5 m (5 ft) long. The African knifefish *Gymnarchus* also can reach 1.5 m (5 ft) in length and is electric. The Nile tilapia (*Oreochromis*), on the other hand, maxes out at 0.6 m (2 ft). The bichirs (*Polypterus*) have lungs and in shallow water breathe through the spiracle behind each eye. All these fish live in a river system up to 10 m (35 ft) deep that discharges on average about 100,000 cubic feet of water per second; the longest river in the world but hardly the largest in terms of volume, depth, or width—a number of rivers are larger by those metrics.

But northern Africa was not always dry with its thin fluvial strip of luxuriant vegetation. Only a few thousand years ago, the vast Sahara was grassland, complete with large mammals we think of today as being more typical of savannas in Kenya and Tanzania to the south. And long before that, in the middle part of the Cretaceous, Egypt was wetter still. In fact, it was a coastal delta setting not unlike that of the Williams Fork Formation in Colorado, discussed earlier in this chapter. About 356 km (220 mi) southwest of Cairo is the Bahariya Oasis and the Bahariya Formation, a middle

Cretaceous unit with 20 varieties of sharks, rays, and sawfish, and more than a dozen types of bony fish, including lungfish, bowfins, coelacanths, pycnodontids, and ichthyodectids. This is a unit worked in recent decades by a joint University of Pennsylvania–Egyptian Geological Museum team and, since about 2016, by Hesham Sallam's Mansoura University crew out of Cairo (fig. 3.9*A* and 3.9*B*). *Bawitius* was a giant polypterid fish, 3 m (10 ft) long, related to the bichirs, an atypical group of endemic fish that today is restricted to freshwater habitats of Africa and the Nile River system. They are singular in having a long row of spined dorsal fins and in having lungs for air breathing, although they are not lungfish, just very interesting basal ray-finned fish.[6] There was also a second, much smaller polypterid in the same deposits. And in the formation, fossil fish generally can be quite plentiful relative to other vertebrates. Certainly, they are diverse. Another large Bahariya fish was the coelacanth *Mawsonia*, which was so large that it takes a full-blown quarry to remove even a partial specimen of it (fig. 3.9*C*–3.9*F*). The aforementioned true lungfish appear to be of several types (fig. 3.9*G*).

The Bahariya Formation is the unit that produced the dinosaurs *Paralatitan* and *Spinosaurus*, the former a relatively recently found giant sauropod. The large theropod *Spinosaurus*, however, was first found in the area in the early twentieth century and, appropriately, appears to have been a specialist in piscivory. The Bahariya Formation fauna includes a moderately diverse set of nondinosaurian reptiles. This unit represents a shallow marine to coastal setting along the southwestern shore of the Tethys Sea. It includes, at different levels and geographical areas, deposits indicating shallow seafloor, barrier islands, beaches, mangrove-convergent tree fern swamps (also with crabs and ferns among other fossils), lagoons, oyster reefs, fluvial channels, and tidal flats and channels.

In the desert west of Luxor and well south of Bahariya lies the Kharga Oasis and the Campanian-age Quseir Formation, another shallow marine to freshwater river and lake deposit, though not as well known. This formation has produced lungfish (as has the Williams Fork Formation), a crocodilian, and the sauropod dinosaur *Mansourasaurus*.

Among the last of the fish to live with dinosaurs were sharks and amioids such as the giant *Melvius*, a 3 m (10 ft) bowfin that lived in the rivers and ponds during the Late Cretaceous in the Williams Fork and Hell Creek formations, among others. *Melvius* is known from fragmentary material in the Hell Creek Formation and a few other formations of the Rocky Mountain region, but a partial skull was recently described from the Upper Cretaceous of New Mexico. Also, Jaelyn Eberle of the University of Colorado recently found jaw and skull pieces of a large *Melvius* when we relocated one of the sites that Dave Archibald and us field assistants worked in the Williams Fork decades ago in Colorado.

The Williams Fork, Quseir, and Bahariya Formations are far from the only Late Cretaceous units to produce such a wide variety of fish from coastal swamp and delta deposits, but they are a few of the best represented. The Kaiparowits Formation of southern Utah is similar in age to

A
C
element with
tooth shagreen
left gular
001-13
left
postparietal
also from quarry but not mapped:
001-14 (2 scales)
001-15 (scale)
several large bone fragments
left quadrate
missing
right
squamosal
left
squamosal
left
opercular
001-02
right
postorbital
right pterygoid
complex
l. ectethmoid
parasphenoid
1.5m
2.0m
2.5
3.0m
3.5m
BAHR001
HAPPY FISH HILL
1/21/00 PKV
N
10° EAST
B
LEFT OPERCULAR
LEFT SQUAMOSAL
LEFT POSTPARIETAL
LEFT GULAR
D
GEOLOGY AT PENN
E
F
rostral process
projection
G

***Facing,* 3.9** Scenes from the Upper Cretaceous of Egypt, hunting for dinosaurs and beast companions in the Bahariya Formation. (*A*) Hesham Sallam of Mansoura University in Cairo poses in an arm's-length photo with his students and crew in 2018. (*B*) Fieldwork in 2016, the Mansoura crew at a fish quarry that produced the giant fish *Bawitius*. (*C, D*) Map and photo of another quarry producing the giant coelacanth fish *Mawsonia*. (*E*) Gular (skull-throat bone) of *Mawsonia* in quarry shown in *C* and *D* (2005). (*F*) Postparietal (skull bone) of *Mawsonia* in same quarry. Scales in *E* and *F* in centimeters. (*G*) Lungfish tooth plates from the Bahariya Formation. Photos and map by and courtesy of Hesham Sallam, Mansoura University Vertebrate Paleontology Center (*A*), Sanaa El-Sayed, Mansoura University Vertebrate Paleontology Center (*B*), Barbara Grandstaff (*C, D, F,* and *G*), Allison Tumarkin-Deratzian (*E*), and Josh Smith (*F*), by way of Matt Lamanna.

the Williams Fork, but it represents a slightly more upland paleoenvironmental setting farther inland from the coast. The Kaiparowits nevertheless includes a comparable fish fauna: gars, pycnodontids, semionotids, amioids (including *Melvius*), and numerous teleosts, plus sharks, rays, and guitarfish. Meanwhile, about 15° latitude farther north in what is now Alberta, the bony fish fauna of the contemporary Dinosaur Park Formation consisted of many of the same groups and genera but lacked gars and some semionotids, pycnodontids, and amioids. It also features several additional forms, including sturgeon, that have not yet been found in the Kaiparowits. Additionally, the relative abundance of common teleost taxa appears to vary between the two latitudinal positions.

First Vertebrates

As the first vertebrates the world saw, back in the Cambrian, it is not surprising that fish are common in almost all marine and freshwater settings in Mesozoic rocks. Their ubiquity and diversity throughout the Age of Dinosaurs is a reminder that ancient watery environments were as teeming with life as anything we see today, and many of the same groups made those waters home even if the relative abundance of more primitive and more derived forms have shifted somewhat since the middle of the Mesozoic.

The fish we saw in the Williams Fork Formation in Colorado only hinted at what was there. For all the fish and other fossils we had collected over the summer, the majority had not yet been found. It wasn't until all of the matrix we had washed in the cattle tanks that summer was back in the lab and the concentrate picked under a microscope that all of the tiny teeth and scales were noticed and counted. Just as an example, over the years, Dave's crews ended up collecting more than 1300 teeth just of the cartilaginous fish. Microvertebrate fossils often require this two-part work of field and lab to find everything that was brought in, which is nice because it often saves a second round of surprises for when we get back to the museum. Such lab work on matrix from micro sites just found in 2012 and 2019 has recently revealed several species of fish and sharks newly recognized in the Williams Fork Formation. More species than we previously realized were present. So the surprises from the Mesozoic fish market continue.

Notes

1. You may notice that we do this a lot.

2. It has held up well as a camp arrangement in the years since.

3. I've since seen people do exactly this, but at age 22, I saw the grad school choice as a now-or-never situation as far as paleo was concerned.

4. This material had probably been acquired by Don's predecessor, William Morris.

5. The Osteichthyes includes two groups of bony fish, the sarcopterygians (flesh- or lobe-finned fish such as lungfish and coelacanths) and the actinopterygians (ray-finned fish such as trout, muskies, tuna, and all the rest). The first tetrapods (terrestrial vertebrates) evolved from among a few sarcopterygians making their way onto land. Thus, all of us terrestrial vertebrates are technically also members of Osteichthyes.

6. Lungs are actually common among all bony fish, but they are modified into buoyancy organs called swim bladders in most species.

Smooth Amphibians

Frogs, Salamanders, and Cohorts

4

IN THE NORTHERN PART of Dinosaur National Monument in northeastern Utah, not far from the famous quarry wall with its 1500 exposed Late Jurassic dinosaur bones, there is a small-room-sized quarry near the top of a steep mudstone hill in layers of Brushy Basin Member of the Morrison Formation. This site produced the salamander *Iridotriton* and the frogs *Rhadinosteus* and *Enneabatrachus*, the first known from a partial skeleton and the second from more than a half dozen partial skeletons on one slab. Then there is the matrix, which yields dozens of isolated amphibian bones. The site is Amphibian Central in a formation otherwise not exactly known for its amphibian record. It is truly one of the only sites in the entire formation that can be considered amphibian rich. But amphibian material is rare in the Morrison Formation, with just a few dozen, mostly isolated, specimens found at other sites so far. These rare skeletons at Dinosaur National Monument are thus particularly important. Salamanders and frogs in the Morrison are understandably overshadowed in most peoples' minds by the giant sauropod, theropod, and stegosaur dinosaurs of the Late Jurassic in North America, but in certain situations, they can tell us more about a locality and the formation than the giants themselves.

There was a time when amphibians were not small specialists in eating insects and other small invertebrates, as we know today from frogs and salamanders. No, during the Late Triassic and earlier, there were stem amphibians known as temnospondyls (including metoposaurs) with flat, shovel-shaped heads and bodies up to 4 m (13 ft) long. These semiaquatic ambush predators had conical teeth the same size as those in alligators and were probably nearly as dangerous to those venturing in and near the water. The temnospondyls, while taking a bit of a hit at the Triassic–Jurassic boundary, hung on, at least in Asia, into the Middle to Late Jurassic. But by then, the amphibians of the world were mostly descendants of a small salamander-like ancestor and were becoming the forms we know today: hopping frogs; squirming salamanders and albanerpetontids; and slithering, legless caecilians. In fact, most of these lissamphibians ("smooth amphibians"), as the modern ones are called,[1] appeared earlier, but by the Middle Jurassic, they were nearly the only amphibians left, the glory days of giant metoposaurs being long over by then.

Amphibians appeared in the Paleozoic era long before the Age of Dinosaurs, and the ancestors of modern lissamphibians were likely on the scene before the dinosaurs arrived. This is based in part on molecular studies of likely divergence times of the modern groups and on the fact that a possible early caecilian fossil appears alongside *Coelophysis* in the Late Triassic, suggesting that the three groups had diverged before then.

Lissamphibians comprise the modern salamanders, frogs and toads, caecilians, and the recently extinct salamander-like albanerpetontids, a separate family with representatives ranging from the Middle Jurassic to the Pliocene, just before the recent ice ages began in the early Pleistocene. Lissamphibians effectively are derived temnospondyls, having descended from a salamander-like Paleozoic ancestor similar to little *Doleserpeton* from the Permian of Oklahoma. This small amphibian with body, limbs, tail, and skull similar to modern salamanders was only a few inches long and appears to have been significantly terrestrial but with a reproductive dependence on water. From forms similar to this evolved not just the similar salamanders and newts of today—plus albanerpetontids—but also frogs and caecilians.

Hoppers: Frogs

Frogs today comprise more than 6000 species, and they make up more than three-quarters of all amphibian diversity. They are specialized amphibians with no tail as adults, extralong hind limbs with relatively large feet for jumping, a pelvis with a bit of lever action to add energy to jumps, a short trunk, and a relatively large skull. In addition to excelling at jumping, some frogs swim, walk, or even glide between trees, as in the gliding tree frog of Central and South America (*Agalychnis spurrelli*). Although the vast majority of frogs, and lissamphibians in general, are freshwater, a few can tolerate life in brackish water. Frogs develop from egg-hatched aquatic tadpoles, as seen on school trips the world over, and this metamorphosis is more pronounced than what is seen in salamanders or caecilians. Some species, however, retain the eggs in the stomach or a pouch until they have hatched or longer and then release tadpoles or fully developed froglets from the mouth. Fossil tadpoles from the Early Cretaceous indicate that these larval frogs were, during the Mesozoic, essentially of the same proportions as we know today, but then, many Cretaceous frogs are classified among the crown (still living) groups, so this might be expected. At any rate, Mesozoic amphibians were likely no less impressive in their ecologies than modern ones, and thankfully, many more fossils of these delicate amphibians have survived to be known to scientists today.

South of Tuba City, in northern Arizona, is a long outcrop formed of Ward Terrace and the Adeii Eichii Cliffs, home of what is known as the "silty facies" of the Kayenta Formation. This is remote territory in and near the Navajo Nation, running from south of town to the southeast toward the Painted Desert, a land awash in pastels and starker shades—one that I can't help but think of as "Mary Colter terrain."[2] It is a land of spectacularly red sandstones and mudstones, hoodoos (rock columns), and dry streambeds where very little grows, but its landscape is dotted with localities producing Jurassic dinosaur tracks, dinosaurs, and other critters. The Kayenta Formation is an Early Jurassic–age deposit where crews from Harvard and Northern Arizona Universities worked in the early 1980s. The formation, and especially the Gold Springs Quarry, has produced some of the earliest

4.1 Some fossil frogs of the Mesozoic. (*A*) *Gracilibatrachus* from the Early Cretaceous–age Las Hoyas deposits of Spain. (*B*) *Genibatrachus* from the Early Cretaceous–age Longjiang Formation of China (IVPP V24104) next to a shell of a conchostracan. (*C*) The spade-foot toad (pelobatid) relative *Liaobatrachus grabaui* from the Early Cretaceous–age Yixian Formation of China (IVPP V11525). (*D*) *Liaobatrachus zhaoi* also from the Yixian Formation (IVPP 14979.1). All scale bars in centimeters. *Image in A courtesy of Angela Delgado-Buscalioni; B–D courtesy of Liping Dong, Wei Gao, and the IVPP.*

4.2 Life at a Late Jurassic pond in what will become the Morrison Formation of southeastern Utah. A frog floats at the surface scattered with leaves of ginkgophytes while below the surface, a pair of amioid (bowfin) fish investigate the frog and vomit up a recent meal of frogs, respectively. *Artwork by Brian Engh.*

of the more modern turtles, mammals, and other small animals in North America. We will discuss the faunal members of this Kayenta biota again and again in these chapters. Among those small Kayenta animals is an early frog, one of the first to appear on the continent.

Frogs, more specifically the Anura, which technically includes both frogs and "toads" (the latter of which is not a natural group), appeared in at least a very early iteration in *Triadobatrachus* in the early Triassic of Madagascar. This animal had a relatively long trunk and hind limbs that were not quite as elongate and specialized as in later forms, so although it appears to have been a lissamphibian, it is well outside the radiation of modern frogs and thus is a stem frog. But frogs do appear in North America by the Late Triassic in the Chinle Formation of Arizona, fossils of an as yet unnamed taxon. And by the Early to Middle Jurassic, there are also the stem frog *Prosalirus* from the Kayenta Formation of Arizona, mentioned previously, and *Vieraella* from the Roca Blanca Formation of Argentina (Early Jurassic) and *Eodiscoglossus* from England (Middle Jurassic). *Prosalirus* lived alongside the carnivorous dinosaur *Dilophosaurus*, minus the latter's imaginary Hollywood venom and neck frill. Late Jurassic to Early Cretaceous frogs include the several forms from the Morrison Formation (e.g., *Rhadinosteus*, *Enneabatrachus*) plus taxa from Argentina (*Notobatrachus*), Spain (*Gracilibatrachus* [fig. 4.1A]; *Eodiscoglossus*), Portugal, China (e.g., *Genibatrachus* and *Liaobatrachus* [fig. 4.1*B*–4.1*D*]), and the UK. By this point, frogs had achieved the specialized, long-legged skeletal morphology and hopping ability we know and love them for today. Skull morphology was also generally similar, and these Late Jurassic to Early Cretaceous forms were within the crown radiation of frogs, so ponds of the Jurassic were likely sites of the summertime frog choruses so familiar to us today.

The Jurassic ponds also appear to have been the sites of occasional ambushes of frogs by aquatic predators of the time. A site in the Morrison Formation in southeastern Utah, one otherwise very productive for fossil plants, yielded what appears to be the regurgitated remains of several tiny frogs. It is not clear what predator ate the frogs, but there are fossils of bowfin fish at the site too, so these are a good possible suspect (fig. 4.2).

By the later part of the Cretaceous, frogs had been living alongside dinosaurs for some time, and finding their bones (such as *Palaeobatrachus*) mixed in with the microsites of the Upper Cretaceous is common even if the remains aren't superabundant. Frogs are more widespread by this point as well, with forms known from Europe, Asia, Africa, South America, and North America. The Upper Cretaceous Maevarano Formation of Madagascar contains the bones of *Beezlebufo*, a frog with a large head that grew potentially up to 42 cm (16 in) long and that was likely capable of eating very young dinosaurs. Clearly, Mesozoic frogs were not primitive in either diversity or ecological range.

Squirmers: Salamanders

Like frogs, salamanders probably also evolved from a late Paleozoic form similar to *Doleserpeton*, and they appeared in the Late Triassic in the stem form *Triassurus* from the Madygen Formation of Kyrgyzstan. As with other lissamphibians in the Mesozoic, there is a bit of a gap from this taxon to better-known salamanders in the middle part of the Jurassic and into the Late Jurassic. By these times, the world had welcomed salamanders such as *Egoria* from the Middle Jurassic of Russia, *Pangerpeton* from the Middle Jurassic of China, *Marmorerpeton* from the Middle Jurassic of England, *Jeholotriton* and *Beiyanerpeton* from the Middle–Late Jurassic of China, the stem form *Karaurus* from the Middle to Late Jurassic of Kazakhstan, and the possible crown form *Iridotriton* from the Late Jurassic of North America.

The Middle–Late Jurassic salamander forms *Jeholotriton* and *Chunerpeton* from China (fig. 4.3A and 4.3*B*) have been found, based on stomach contents, to have specialized in eating conchostracan arthropods and nepomorph hemipteran insects, respectively, a window into Mesozoic salamander diets provided by rare soft-bodied preservation of these amphibian taxa. Modern salamanders feed on insects and insect larvae, small crustaceans, snails and slugs, worms, and fish (in more aquatic taxa). Middle Jurassic salamanders, such as *Neimengtriton* (fig. 4.3C), are also sometimes well preserved.

4.3 Some fossil salamanders of the Mesozoic. (*A*) *Jeholotriton* with stomach contents of conchostracans from the Middle–Late Jurassic–age Yanliao Biota near Daohugou, China (IVPP V14195). (*B*) *Chunerpeton* from the same locality and age as *A* (IVPP V13343). (*C*) *Neimengtriton* from the Middle Jurassic Haifanggou Formation of China (IVPP 13393). *All images courtesy of Liping Dong, Wei Gao, and the IVPP.*

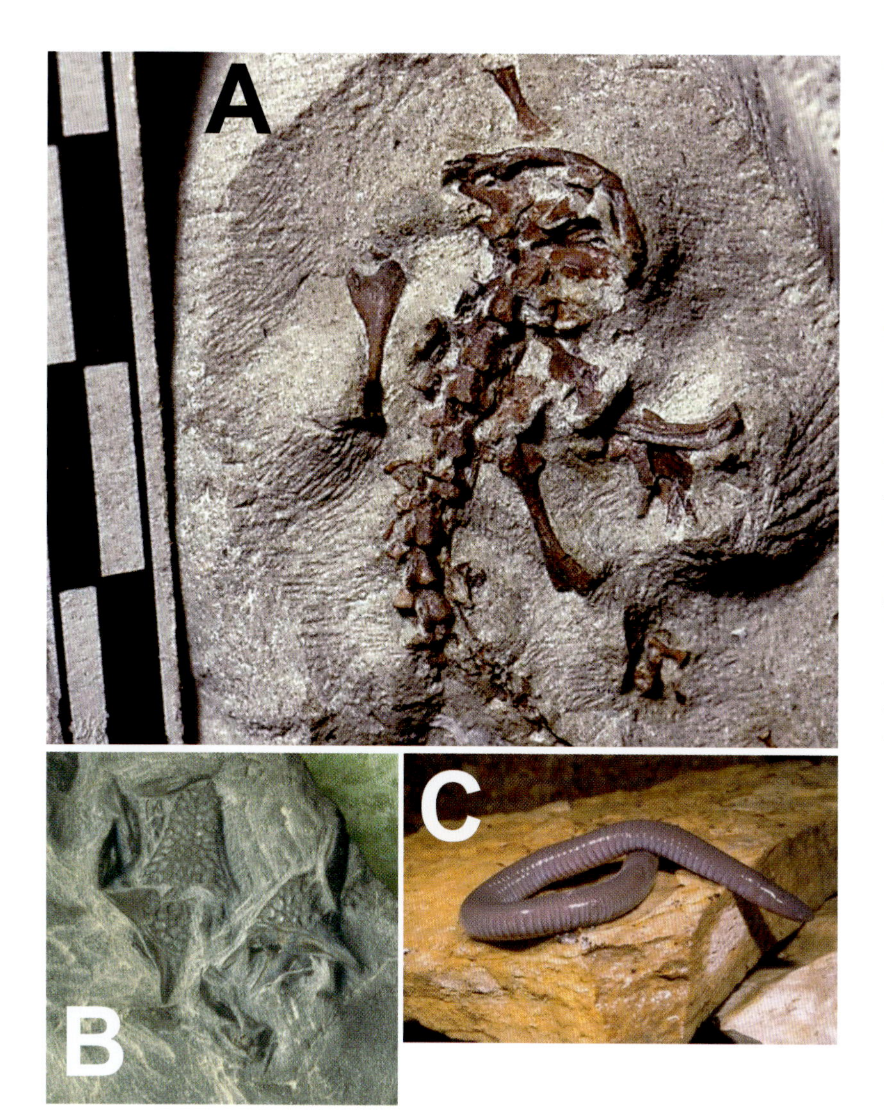

4.4 Some Mesozoic and modern amphibians. (*A*) The Late Jurassic salamander *Iridotriton* from the Morrison Formation of Utah (DINO 16453). (*B*) Elements of the albanerpetontid *Shirerpeton* from the Early Cretaceous of Japan (SBEI 12459). (*C*) The modern caecilian *Dermophis mexicanus* from Mexico and Central America. *Photo in A by author, B courtesy of Susan Evans and © Shiramine Institute of Paleontology, C by Franco Andreone, from Wikimedia Commons (https://commons.wikimedia.org/wiki/File:Dermophis_mexicanus.jpg).*

Iridotriton is a small, nearly complete salamander skeleton from the Morrison Formation at Rainbow Park, from the north end of Dinosaur National Monument, Utah (fig. 4.4A). The limbs, vertebrae, and skull are in general very similar to salamanders today, so there is little "ancient" or "primitive" about this taxon, or most other Mesozoic salamanders, for that matter. Stepping out of a time machine, we would likely have had no problem identifying *Iridotriton* as a salamander in its natural environment. There are hints in the Morrison Formation, in the form of isolated or fragmentary bones from Dinosaur National Monument, Utah, and Como Bluff, Wyoming, that North America may have also been home to karaurid salamanders during the Late Jurassic.

All salamanders are (and were) dependent on water for reproduction, as are most lissamphibians, and today, they inhabit ponds, wet forest floors, and other moist areas of the landscape. Impressively, salamanders today can regenerate entire limbs after injury, and Mesozoic forms may well have been able to do the same. The more than 650 species of salamanders around today include forms that go through a minor metamorphosis after hatching and have distinct larval and adult morphologies as well as others that retain juvenile characteristics as adults (neotenic). Salamander larvae have an elongate trunk, a finlike tail, and rather obvious external gills; adults of the nonneotenic species lose the gills and caudal fin on the tail. Neotenic salamanders are generally more aquatic. In Mexico, for example, the endangered axolotl (*Ambystoma mexicanum*) retains its gills and tail fin and lives in the canals that are all that remain of the series of large lakes that once existed where Mexico City now stands—hundreds of years ago, the axolotls were likely quite numerous in these lakes. In much of Mexico and the United States, meanwhile, the wide-ranging North American tiger salamander (*Ambystoma tigrinum*) lives mostly in terrestrial environments and only returns to water for reproduction. Terrestrial salamanders that lose their gills will either develop lungs or will breathe through their skin, depending on the species. And it appears that ancient salamanders were as developmentally diverse. The Jurassic *Jeholotriton* from China appears to have been a neotenic, aquatic salamander.

Squirmers Reprise: Albanerpetontids

Albanerpetontids have a generally salamander-like body plan, and for some time, they were thought to be salamanders, but they may in fact be similar to the common ancestor of lissamphibians, or they may be a sister taxon to salamanders. One of the main anatomical differences between albanerpetontids and other lissamphibians is that the former lack the smooth skin of the others and instead have skin covered in small, almost fish-type scales (scales are present, however, in most caecilians too). There are only five genera of albanerpetontids, with most of them only named in the past 25 years. The taxa range from *Anoualerpeton* in the Middle Jurassic Kilmaluag and Forest Marble Formations of Scotland and England, respectively, plus the Early Cretaceous of Morocco; *Celtedens* from the Late Jurassic of Portugal, the Early Cretaceous of the UK, and the Late Cretaceous of Spain; *Shirerpeton* (fig. 4.4*B*) and *Wesserpeton* from the Early Cretaceous of Japan and England, respectively; and *Albanerpeton* is known from the middle of the Cretaceous through the latest part of the period, and then from the Paleocene, Miocene, and Pliocene, all almost entirely from North America or Europe.

Albanerpetontids have teeth and jaws similar to those of salamanders, so they may well have had a similar diet of insects and other small invertebrates. As with other lissamphibians, they were probably restricted to remaining close to sources of water for reproduction.

Slitherin': Caecilians

Caecilians are elongate, limbless amphibians that live today in the tropics in most parts of the world, except Madagascar and many other ocean islands. To some extent, they resemble giant (though vertebrate) earthworms (fig.

4.4C). Most burrow through soft, wet soil in proximity to rivers, swamps, or lakes, although some species are aquatic. To assist a burrowing lifestyle, the eyes are reduced in size and covered by skin, and the skull is heavily ossified.

Caecilians likely originated from the same salamander-like group as frogs and salamanders and are within the Gymnophiona, which comprises Caecilians and stem-group taxa (including all Mesozoic forms known so far). Their fossils are rare. *Rubricacaecilia* is known from the Early Cretaceous of Morocco. Fragmentary remains are known from the Late Cretaceous of Sudan and Bolivia and the Paleocene of Brazil.

Caecilians may have appeared as early as the Late Triassic in the form of *Chinlestegophis* from Colorado outcrops of the Chinle Formation. But at least by the Early Jurassic, there was the stem form *Eocaecilia* from the "silty facies" of the Kayenta Formation of Arizona, a genus that, unlike modern caecilians, retained small fore and hind limbs.

Lissamphibian Precision

During that long summer collecting anthills in the Williams Fork Formation in northwestern Colorado, we did not find any amphibians during our picking and screenwashing, but they were there—eventually found by one of Dave's crews in an unknown year (maybe it was us?). Of the four lissamphibian groups, it appears that only a probable *Albanerpeton* allowed itself to be found. It likely only showed up in San Diego during the picking of the screenwashing concentrate. All the groups were probably present 70 million years ago in this area, but their fossils do tend to be pretty rare. In 30 years of working in the Morrison Formation, I've found a grand total of five amphibian specimens: a frog femur in Utah (just last year), a partial probable frog humerus and salamander vertebrae from Wyoming, and a possible salamander femur and other elements (in one pile) from Utah. Again, not the most common fossils.

Modern amphibians are another group with origins dating back to before the dinosaurs (though possibly just before). After the amphibian late Paleozoic heyday, and after remaining large and dangerous through most of the Triassic, at least among the metoposaurs, the lissamphibians have kept a low profile in ecosystems from the Jurassic through today. From the lissamphibians' origins in the late Paleozoic or Early Triassic, the frogs, salamanders, caecilians, and albanerpetontids have been a small component of Mesozoic fossil assemblages, but this is more likely due to their small size and the very delicate nature of their bones. They are simply less likely to fossilize. Although amphibians are rarely among the most abundant animals in many ecosystems today (despite the occasional frog chorus or salamander horde around a pond), they probably were equally common in dinosaur landscapes as they are now—certainly in greater proportion than we are seeing in the Mesozoic fossil record.

Amphibians help serve as environmental indicators for ancient environments thanks to their requirements remaining more or less unchanged throughout their history. Frogs and salamanders can demonstrate a variety of ecologies today, of course, but few species tolerate very arid conditions, and nearly all are dependent on water for reproduction. So finding amphibians in a formation otherwise stuffed with dinosaurs gives a more precise indication of local environments than one would get by finding, say, a sauropod dinosaur that could range across environments and migrate hundreds of miles, then end up with its skeleton buried in nearly any environment. Amphibians are one of those groups that help give us smaller data points in the fossil record, windows into the small-scale environments at particular times and particular places in biologic history that eventually allow us to stitch together a more detailed and broadly accurate picture of dinosaur ecosystems.

Notes

1. "Smooth amphibians" are those with smooth glandular skin, in contrast to the scaly skin of other herps like lizards and snakes (chap. 7).

2. Although there are effectively no buildings in the Ward Terrace area, it is almost surrounded by Colter's architectural works.

Mysteries Dark and Vast

Turtles

5

AFTER LIVING FOR NEARLY five years in Boulder, Colorado, I was less than enthusiastic about having to leave town. I had gotten used to the setting, the sports, and the ease of getting around by bike, foot, and bus, and I was, you might say, fully acclimated. While it's not exactly inexpensive in Boulder, there are worse places to be stuck for half a decade while studying the Earth. But I was finishing my dissertation in geology at the University of Colorado and would soon be done. I needed work. Luckily, just before I finished and found myself with nothing to do, I got a call from David Gillette, state paleontologist of Utah at the time, and before I knew it, I had something to do as soon as I was finished, and that was soon.

Grand Staircase

Dave needed help with the Utah Geological Survey's paleontological inventory of the newly established Grand Staircase-Escalante National Monument in southern Utah. The one-year survey project was just about half over, and he needed a new field lead, so within days of graduating, I was off to Salt Lake City and then the southern Utah town of Tropic, east of Bryce Canyon National Park. Grand Staircase was so expansive; it extended over three physiographic provinces from the Staircase itself in the west outside Kanab, across the Kaiparowits Plateau southeast of Tropic, across the Escalante Canyons area and into the Circle Cliffs, bordering Capitol Reef National Park. At 1.8 million acres and with rocks ranging from Permian to Cenozoic, it was far more area and more geological formations (stratigraphic units) than could reasonably be covered in just a year. The project went on long enough that before the full year was over, my predecessors, paleontologists Alan Titus and Gus Winterfeld, had to leave to take more permanent gigs before it was complete. Only our field assistant Josh Smith lasted for the whole project and put up with all three of us. But the full year wasn't nearly enough time to cover the entire acreage of the monument, and we didn't pretend it was. *Outside* magazine had once suggested that a point on the Kaiparowits Plateau was the most remote point in the lower 48 states (defined as the point farthest in all directions from any paved road), and this was probably true—there were entire topographic quadrangles in the Grand Staircase[1] that we couldn't even get into without driving all day and then hiking many miles of roadless terrain—and we didn't get to a number of these quadrangles at all. The paleontological resources we were inventorying were so abundant that we could rarely get more than half a mile from where we had parked our Bronco on some dirt road before we had sites we needed to document. Had we stuck to such areas and tried to cover everything, we never would have gotten off more than a few quadrangle maps. Instead, we began adopting the approach of finding areas of exposure of formations that we had relatively fewer sites from or areas where we had few sites at all. There were many quadrangles for which we had no sites because we hadn't gotten there yet, so we started targeting these "empty" quads just to "spread the sites around" a bit. We realized very quickly that covering everything would be virtually impossible.

It wasn't a bad four months of fieldwork, before the final two months of data processing. Living out of Tropic, Escalante, or Kanab in southern Utah, we'd load into the Bronco in the morning, pick an unexplored quad on which to access a particular formation, and head out, sometimes needing to drive over three hours just to get where we could start looking. If the driver got the Bronco stuck in ruts, mud, or snow, he had about ten minutes to get unstuck before the passenger got a shot at unsticking us. Then it was a process of hiking for hours looking for bones and other fossils and logging what we found, often within only a few minutes. We found logs, footprints, bones, shells, teeth, and leaves, and in many units.

One night, while living in Escalante, we went to J. Howard Hutchison's house for a kitchen-table poker game that I bombed as usual, but at least I didn't completely run out of money until nearly the end of the session. Howard retired to Escalante and is an expert in fossil turtles, and among the fossils we found that fall in the Grand Staircase were a number of his favorite subjects. We found several of these reptiles, usually shell fragments, shells, or limb elements, but nothing spectacular. The more interesting turtles were found by the teams of paleontologists that came soon after. Many of these were freshwater turtles from the Kaiparowits Formation, a Late Cretaceous unit exposed rather extensively, often as steep badlands on the plateau of the same name. The formation has at least 14 species of turtles known from it, and this is not atypical for Late Cretaceous formations in western North America.

Some of the better-preserved turtles found in the Kaiparowits Formation were collected recently by the Denver Museum of Nature and Science. Early in 2017, I was in Denver to give a talk at the museum, and my daughter and I got a tour of its new behind-the-scenes prep lab from their lab manager and field coordinator, Mike Getty. I'd known Mike for almost 17 years by that point, ever since he'd been working the Cleveland-Lloyd Quarry (in the Morrison Formation) for the University of Utah when I visited the site in 2001. Mike had a talent for field cooking, although he must have had a cast-iron stomach. My introduction to his serving skill was tasty sandwiches he made for us all once while tailgating before a reggae show outside Denver that first summer. After that, he managed to make some of the best fry-bread tacos I've ever tasted, with six pounds of ground beef and some moldy salsa (among other ingredients), while in camp working on a sauropod dinosaur in an off-and-on snowstorm in Arches National Park one spring. And no one got sick! On that same trip, he tagged our field truck with plaster while we were away in the hills, forcing us to drive home with "KIRKLAND WAS HERE" in big, bold plaster-stroke letters on our tailgate, the graffiti referring to mutual friend Jim Kirkland, Utah state paleontologist after Dave Gillette.

But Mike's lab tour in 2017 was a sneak peek at some of the largest and best-preserved turtles I've ever seen, from the Kaiparowits or any other formation, and I thought I detected a hint of pride in Mike's showing us around the nice new prep lab too. Unfortunately, that was the last time I talked to Mike for any meaningful period of time, as we only visited briefly at a meeting in Calgary, about six months later, when we met in the halls. He passed away rather unexpectedly at a dig site just three weeks after that Calgary meeting.

The Kaiparowits turtles in Mike's Denver Museum lab were impressive—gorgeously preserved and big, as the freshwater ones go. They proved to be a group of *Basilemys* from one site in the formation, some with soft tissues preserved and one with eggs in place internally. Finding fossil turtles like this is a real treasure, in part because turtle fossils are so common in so many Mesozoic (and Cenozoic) deposits, but 99% of the time, it seems that all we see are shell fragments. To get a nearly complete animal is a rare treat.

We know turtles today as cute, toothless reptiles with beaks and slow movements that live in a mobile home of bony shell into which they can retract their head and limbs for protection. They have lived like that, more or less as we know them now, since the appearance of genera such as *Proterochersis* during the Late Triassic in Europe and then the Early Jurassic with the form *Kayentachelys* in northern Arizona. Although turtles started off as terrestrial and semiaquatic species, and sea turtles were around by the Cretaceous, true tortoises did not appear until after the nonavian dinosaurs had gone extinct. Most of the more than 300 species of turtles today can retract their head and limbs inside a connected shell composed of a dorsal dome (carapace) and a ventral plate (plastron).

Do-It-Yourself Mobile Homes

Turtles arose around the same time as the dinosaurs, although it appears to have been a gradual process that may have actually begun late in the Paleozoic. As the Triassic world became more dangerous with new types of carnivores, it appears that the ancestors of turtles began armoring themselves. These ancestors acquired a couple of turtle characteristics (nine elongate dorsal vertebrae) early on, and they began forming the protective shells that became standard turtle outfitting.

Curiously, although the shell might appear to be dermal armor, more or less equivalent to the osteoderms in the skin of crocodiles, for example, it is possibly a mix of deeper endochondral bone, part of the main skeleton, along with developmentally dermal bone. It was once thought that the carapace of turtles formed by the fusion of dorsal dermal armor pieces to each other and then to the underlying ribs. But this wouldn't explain an oddity in turtles. One longtime question is how turtles came to be unique among vertebrates in having their shoulder blade (or scapula) under their shell and thus *inside* their rib cage. The rest of us tetrapod vertebrates—feel your shoulder blade right now—have our scapula embedded in muscle *above* our rib cage. If dermal armor fused onto the ribs, how did it not trap the scapula in between? On the other hand, how did the scapula get inside the ribs? Not an easy maneuver, it would seem. This peculiar arrangement in turtles may be explained if the carapace is not actually derived evolutionarily entirely from fused dermal armor. We now have a number of new fossils that illustrate how this process may have taken place. It turns out that the carapace probably formed from the gradual anterior and posterior widening of the ribs until they eventually fused to each other to form the carapace, perhaps with the assistance of second, possibly dermal, ossifications in each. There never was a need to fuse anything to the ribs from above because the shell was mostly *made* of modified ribs. And the scapula probably came to be located inside the rib cage by a change in development so that the ribs grew out from the spinal column without curving ventrally as much as they did previously, until they went outside the scapula as they lengthened during development. A potentially problematic result of the fusion of ribs to each other was the loss of the intercostal muscles between the ribs. In most amniotes, breathing is partly facilitated by trunk muscles expanding and contracting the ribs to open and compress the lungs, but with the stabilization of the trunk in turtles, this was no longer possible, so turtles developed the unique use of paired and opposed muscles inside the shell to create the expansion and contraction.[2] The shell halves, carapace and plastron, are the two main components that protect the turtles when they retract themselves inside their shells, and as we will see, turtles achieved this full-shell stage at least 200 million years ago.

However, the development of the carapace seems to have lagged behind that of the ventral shell, the plastron. The plastron likely formed a step or two ahead of the carapace evolutionarily, largely from the expansion of the gastralia (or "belly ribs") and sternum, along with the inward and posterior expansion of the clavicles (collarbones).

The fossils now available that show this evolutionary development in turtles include three forms that show the anterior-posterior expansion of the dorsal ribs and gastralia: *Eunotosaurus* from the Middle Permian of South Africa and Malawi, *Pappochelys* from the Middle Triassic of Germany, and *Eorhynchochelys* from the Late Triassic of China (fig. 5.1). The next stage is demonstrated by *Odontochelys*, also from the Late Triassic of China. In *Odontochelys*, the dorsal ribs are expanded more but not fused, and they extend more laterally with less ventral curvature. The plastron is also well formed, though not complete, and is ahead of the development of the carapace. The final stage is apparent in *Proganochelys* from the Late Triassic of Germany; this form had a full carapace and plastron. A form of this approximate grade has been reported from the Late Triassic–age Chinle Formation of New Mexico, but whether it is *Proganochelys* or a new taxon (*Chinlechelys*) has been debated. Another iconic feature of the transition to full "turtleness," demonstrated starting with *Eorhynchochelys*, is the reduction and loss of the teeth and the development of a beak.

5.1 The stem turtle *Eorhynchochelys* (SMPP 000016) from the Late Triassic–age Xiaowa Formation of China. *Photo courtesy of Nick Fraser.*

Modern turtles can often retract their skull and limbs inside the shell, for the skull in either a typical or "side-necked" fashion. The ability to fully retract the skull and neck appears to have appeared well after full development of the shell, perhaps as late as the Cretaceous. But the phylogenetic split between turtles that retract their head by bending the neck side to side in a horizontal plane and those that fold their neck vertebrae in a more vertical plane ("hidden-neck") appears to have occurred by late in the Jurassic.

Turtles in Surf and on Turf

Fossil turtles now occur almost everywhere we find dinosaurs. Among the *Dilophosaurus* in the Lower Jurassic Kayenta Formation south of Tuba City, Arizona, is the above-mentioned *Kayentachelys*. By the Middle Jurassic, there were turtles like *Eilenchelys* in the Kilmaluag Formation of Scotland (fig. 5.2A). And among the 1500 Late Jurassic bones on display on the wall of rock at Dinosaur National Monument in Utah, mixed in with *Diplodocus*, *Stegosaurus*, and others, were shells of the turtles *Glyptops* and *Dinochelys*, turtle taxa that occur in many parts of the Morrison Formation (fig. 5.2*B* and 5.2C). The Middle–Late Jurassic–age Shaximiao Formation in China, which is the basis of the Zigong Dinosaur Museum, produced four genera of turtles. As we've seen, the Kaiparowits Formation's turtle fauna, of which the large Denver Museum turtles were part of the sample, consists of at least 14 species. Among these was the bumpy-shelled *Denazinemys* (fig. 5.2*D*). More Cretaceous turtles are known from China (fig. 5.3A) and from the Hell Creek Formation in North America, the latter being where some of the last turtles to live alongside dinosaurs are found (fig. 5.3*B*). Very few terrestrial Mesozoic formations don't have at least a few bits of freshwater turtles in them, the animals that, much like today, lived in ponds and rivers,

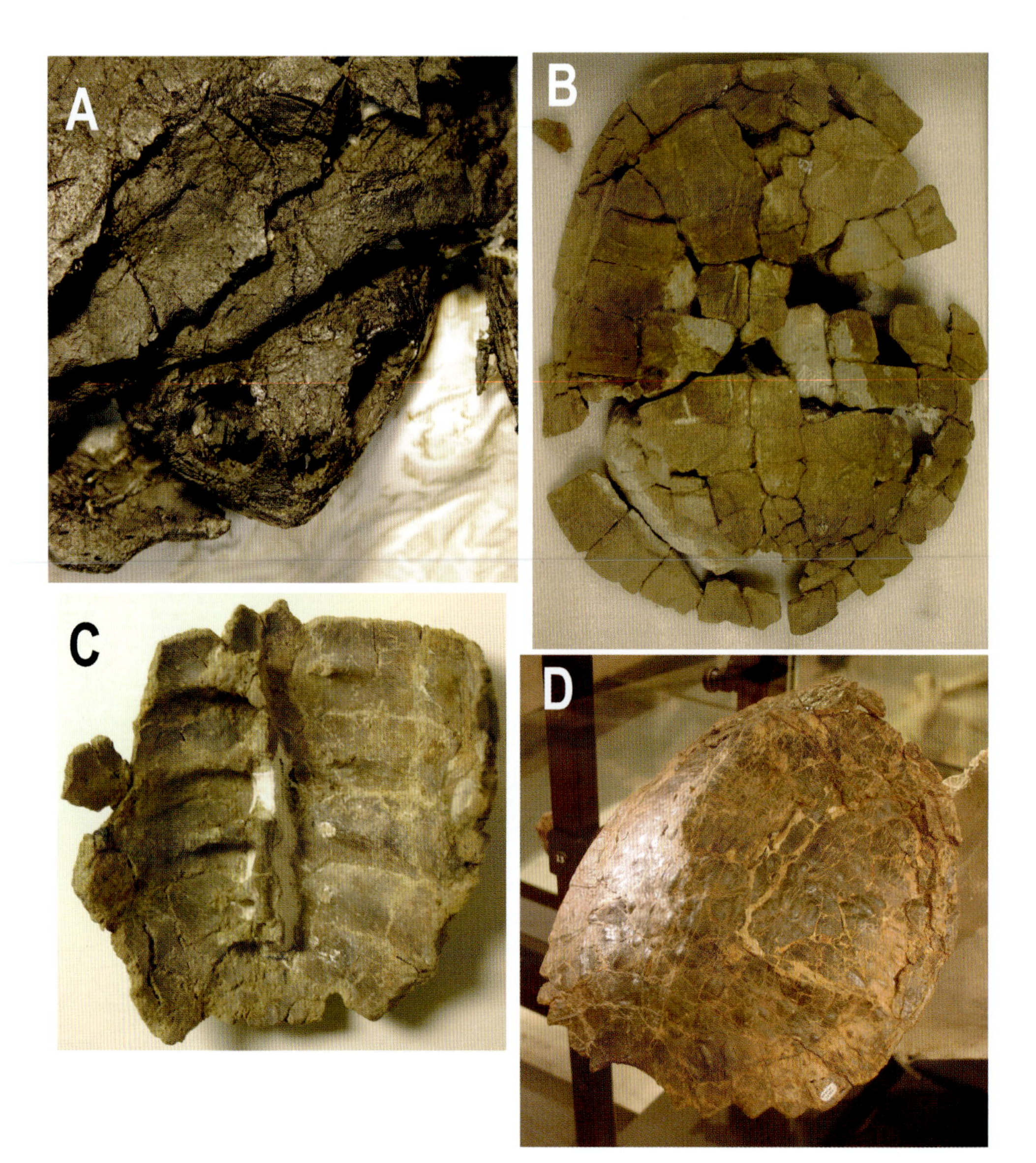

5.2 Some turtles of the Jurassic and Cretaceous. (*A*) Skull peeking out of the shell of *Eilenchelys* (NMS G2004.31.16d) from the Middle Jurassic Kilmaluag Formation of Scotland. (*B*) Partial shell of the turtle *Dinochelys* from the Upper Jurassic Morrison Formation of western Colorado. (*C*) Internal surface of partial shell of *Glyptops* from the Upper Jurassic Morrison Formation of Wyoming. (*D*) Complete shell of the Late Cretaceous turtle *Denazinemys* from the Kaiparowits Formation of southern Utah (UMNH VP20447), as displayed at the Natural History Museum of Utah. *Image in A courtesy of Susan Evans; B–D photos by author, courtesy of the Museums of Western Colorado (B, C) and Natural History Museum of Utah and Bureau of Land Management (D).*

sunning themselves on logs and eating fish and aquatic plants among other elements of a mostly omnivorous diet.

But turtles weren't only in lakes and rivers during the Age of Dinosaurs. In the oceans were sea turtles like the giant *Archelon*, known today from Late Cretaceous–age Pierre Shale deposits in South Dakota. This turtle

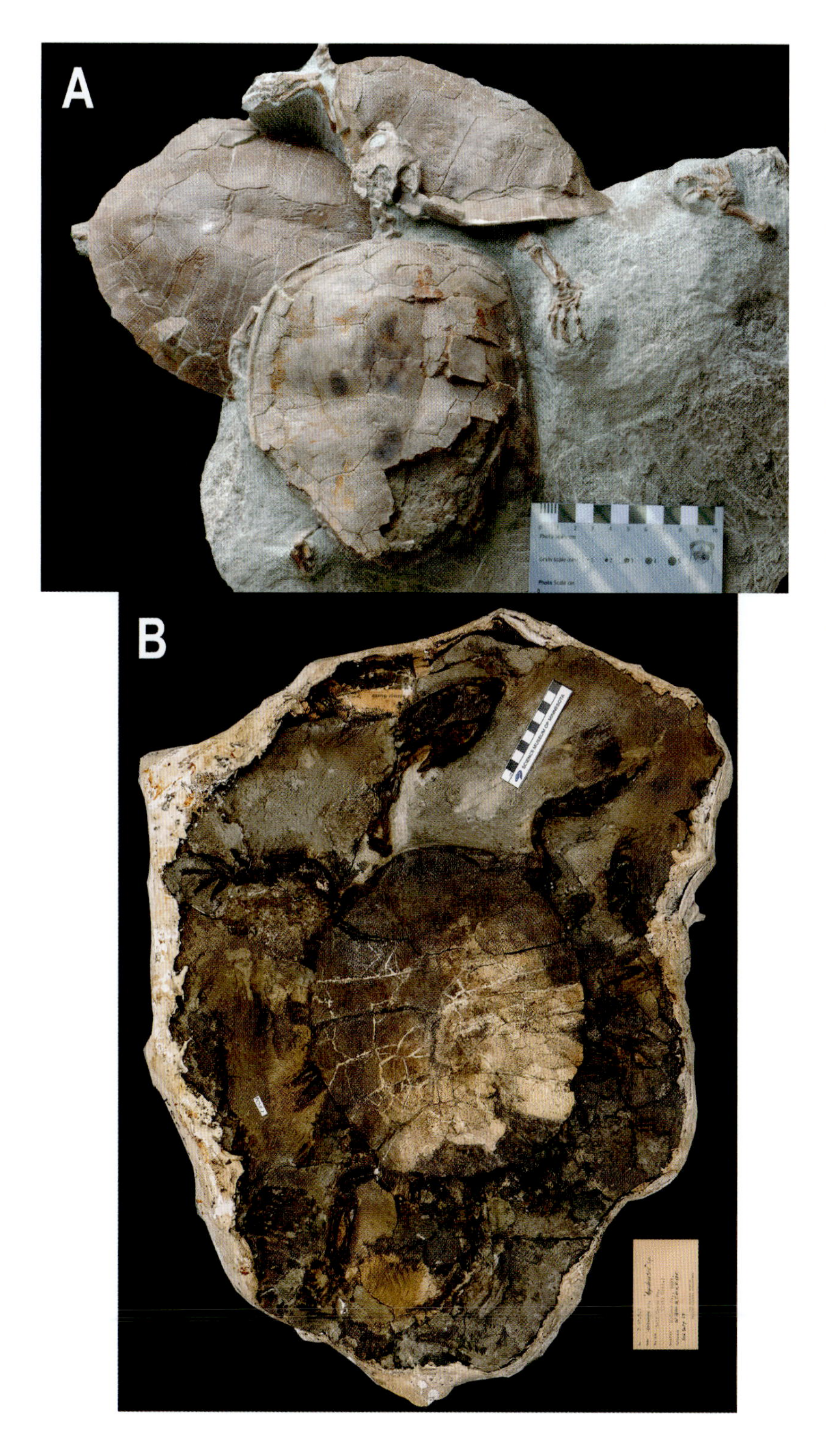

5.3 Some Cretaceous freshwater turtles. (*A*) Three skeletons of the Early Cretaceous turtle *Xinjiangchelys* from China (PMOL-SGP A0100-1, -2, and -3). (*B*) Nearly complete skeleton of the trionychid turtle *Gilmoremys* from the Upper Cretaceous Hell Creek Formation of Montana (SMM P67.5.1). Photo in *A* from Rabi et al. (2013) *BMC Evolutionary Biology* and courtesy of Márton Rabi and Walter G. Joyce. *Photo in B by Mark Ryan and courtesy of Alex Hastings and the Science Museum of Minnesota.*

5.4 Yale paleontologist George Wieland standing next to the Cretaceous *Archelon* sea turtle skeleton he found near the Black Hills. This is a ventral view of the turtle taken in 1914. Wieland is mostly known as a paleobotanist who specialized in fossil cycads, but in addition to this *Archelon*, he also excavated much of the type specimen of the sauropod dinosaur *Barosaurus. Image © Peabody Museum of Natural History, Yale University, New Haven, Connecticut, USA. YPMAR.002076.*

was up to 4.5 m (15 ft) long, head to tail, and 13 ft (4 m) across its paddle-like forelimbs (fig. 5.4). It may have been a predator of smaller food items in the relatively shallow parts of the sea. Encountering one of these while snorkeling the Western Interior Seaway would have been quite an experience. Modern sea turtles are elegant swimmers as it is, but to have an ancient one the size of a VW Bug "flying" toward you through the water would be astonishing.

Archelon was neither the first nor the only sea turtle of the Mesozoic either. Several smaller forms are known from other deposits, including a couple from the Solnhofen Formation of the Late Jurassic of Europe (fig. 5.5). *Thalassemy* and *Tropidemys* predated *Archelon* by a good 80 million years or so.

Turtle fossils occur in seemingly every terrestrial deposit of the Mesozoic, the remains of freshwater animals that existed in ponds and rivers from a time when Earth was warmer at higher latitudes than it is now. Rather than reaching their latitudinal limit in southern Canada (North America) and central Europe, as they do now, turtles ranged to arctic latitudes during the Age of Dinosaurs. Turtles lived all the way up in northern Alaska and

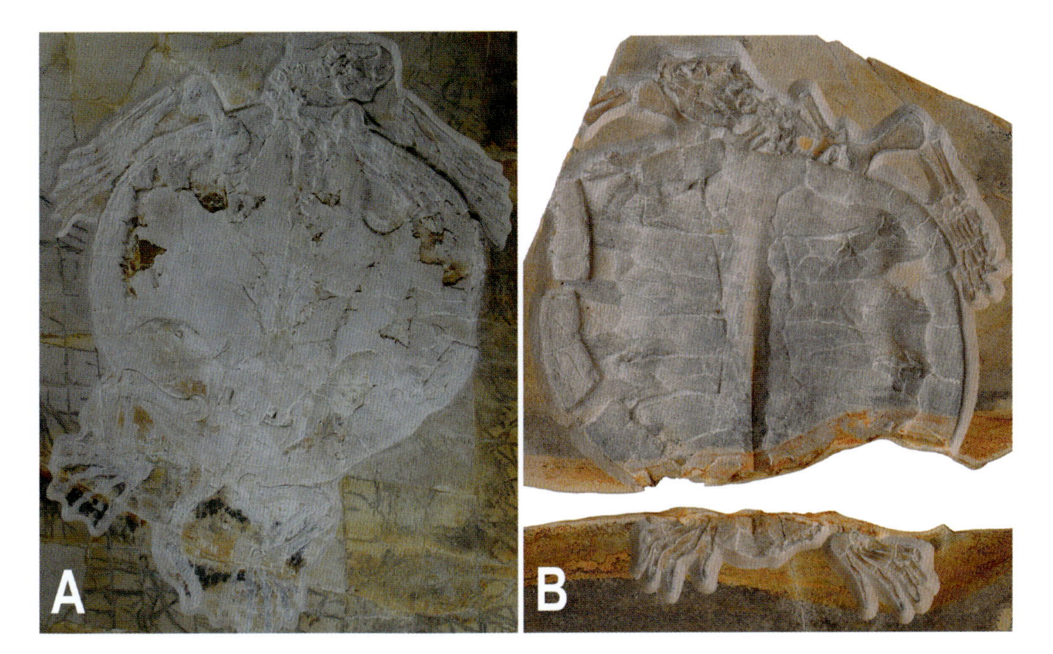

5.5 Aquatically adapted (and possibly marine) turtles from the Late Jurassic of Germany. (*A*) *Thalassemys* (NKMB Watt18/211). (*B*) *Tropidemys* (NKMB Watt09/162). *Photos courtesy of Walter G. Joyce, M. Mäuser, and Serjoscha W. Evers.*

the Canadian Arctic during the Cretaceous, as well as high latitudes to the south. In fact, peak freshwater turtle diversity appears to have moved to higher latitudes throughout the Mesozoic, peaking in the Late Cretaceous. And they had already invaded the sea as well.

Transitions and Origins

As with the lissamphibians, turtles also made their appearance around the same time that dinosaurs evolved. It was once thought that turtles evolved from very archaic reptiles from relatively deep in the later Paleozoic era. This idea suggested that turtles were in fact weird holdovers from the age of the very first reptiles. But that was based mostly on aspects of the skull that now appear to have been unique adaptations that only paralleled the structure of those earliest reptiles. In fact, it appears that the first prototurtles split off from more modern reptiles in the Permian or earliest Triassic and, among modern reptiles, are closest to crocodiles and birds—basically, turtles are likely archosauromorphs, related to the most derived reptiles we know.

The Triassic was an evolutionarily active time. In many ways, it was the birth of the modern world. If the main phyla of animals appeared mostly in the Cambrian period, it was the Triassic that featured the appearance of many of the classes and orders that we see today, at least among the vertebrates. The evolution of the turtle shell is a transition that has come into much better focus in recent years, and it is one of several we will see in this book. The development of that shell took millions of years, but in the big geological picture, it occurred fairly quickly. It seems to have worked well because the turtles of today are not tremendously different from their dinosaur-dodging cousins. Anything modern turtles can do, Mesozoic turtles seemingly could do too, long ago (much better?). They're the toughest.

Notes

1. Each quadrangle, in the mid-latitudes of North America, is about 6.5 × 8.5 mi (10.5 × 13.8 km) across.
2. These muscles are essentially limb muscles with some additional jobs.

6

Beak-Heads

Ancestry of the Tuatara

ITS NAME NOW IS *Theretairus*. Of course, it didn't know that's what its name would be some 150 million years after its time—a name assigned by a single great ape among billions of descendants of an anonymous mammalian contemporary somewhere in the world in its time. Indeed, all it knew during its lifetime was that it inhabited a world of greenery, insects, other scaly animals and a few furry ones, and then the occasional ground-shaking, sun-blocking behemoth that appeared to weigh several million times more than it did. It wasn't concerned with names, its own or anyone else's, because it spent most of its days and years preoccupied with food, shelter, and, now and then, protecting eggs.

In the case of our then-nameless small reptile friend, its food was probably insects and their larvae and the occasional small carcass of a vertebrate. Its shelter was likely a thicket of cycadophyte brush or ferns or possibly a small burrow. And its eggs were something it may have taken a while to develop, possibly reproducing on average only about every four years.

Its time and place were collectively one of the most famous in geological history: the Late Jurassic of western North America, a combination that resulted in today's colorful mudstones and sandstones of the Morrison Formation of Wyoming and other western states. But 150 million years ago, during this still-nameless reptile's time, there was no modern Wyoming geography, climate, or environment. Forget Little America and icy pileups on Interstate 80, forget grass or any flowering plants, forget mountains (at least in this area), and forget nine months of winter and snow. Our reptile inhabited a broad floodplain that stretched from what is now western Utah to the Great Plains and from the middle of New Mexico nearly to Canada. Essentially, the entire area that is now the Rocky Mountain region in the United States was then flat; dotted with ponds, lakes, and wetlands; and crossed by perennial and ephemeral streams. The floodplain was covered with ferns, cycads, ginkgoes, giant araucarian conifers, tree ferns, seed ferns, horsetails, and mosses—almost everything except flowering plants, which had not yet evolved. And oddly for Wyoming, it almost never froze, much less got coated in snow or ice; this was an environment in a world quite possibly devoid of polar ice caps, one with what we today consider tropical animals living at high latitudes north and south, and one in which the current fire-and-ice climate of the Western Interior was millions of years off—at this time, it was hot in the summer and warm and wet in the winter. Almost every physical aspect that defines the Equality State that we know and love today was close to its opposite during the Late Jurassic: low elevation, not high; warm, not cold; wet, not dry; well vegetated, not a wasteland of sage plains (separating gorgeous mountains).

Climatically, it was probably an easier place then to be a small, nameless reptile, if not for some of the other inhabitants of the time. Sure, the Morrison Formation has produced some famous dinosaurs such as *Brontosaurus* and *Brachiosaurus*, but they were gentle-giant herbivores. And *Allosaurus* was too busy tackling, and crushing the trachea of, midsize prey dinosaurs to bother with a tiny reptile. But there was a diversity of small carnivorous theropod dinosaurs who almost certainly liked small prey items—theropods like *Ornitholestes, Coelurus*, or *Hesperornithoides*, for example. These and other predators probably made hiding in those bushes, trees, or burrows a well-honed skill in our small reptilian friend.

The members of its species likely existed in the area for several million years, but we don't know what happened to them because they disappear from the rock record and the Morrison Formation at its upper stratigraphic limits. But during the seven million years recorded by that rock unit, there were undoubtedly millions of individuals of the species that existed, and all it took was one individual's lower jaw to be buried in a pond next to a stream to preserve the group for the future.

One hundred and fifty million years later, it was probably a member of digger Bill Reed's team of excavators that found that jaw in Morrison Formation rock in Wyoming sometime between 1879 and 1889 and sent it back to O. C. Marsh at Yale. But it went unnoticed until 1926 when a paleontologist named George Gaylord Simpson described the new reptile and gave it the official fossil species name *Theretairus antiquus*, the "ancient wild beast companion," in reference to its having been found along with a number of Jurassic mammals. The Greek word *therium*, meaning roughly "wild beast," is a common root ending for fossil mammals throughout the Mesozoic and Cenozoic eras of geological time, so Simpson named the new little reptile for its co-occurrence with the furry set of the Morrison Formation's small animals.[1] So now the nameless reptile was officially named. In fact, all members of its species were collectively named. *Theretairus* was in a sense resurrected by its jaw and its recognition in the history of the Earth.

Perhaps most alluring, however, was the type of reptile *Theretairus* was. As a rhynchocephalian, it was a relative of the modern tuatara *Sphenodon*. The fact that Simpson named and described a sphenodontian is itself interesting because, although he is well known as a prominent evolutionary biologist of the midtwentieth century, most of his work at the time was on the Mesozoic mammals of Europe and North America, and he published classic monographs on them in 1928 and 1929. But the importance of a second rhynchocephalian in the Morrison Formation is apparent in that Simpson recognized and worked on this new species of reptile.

Rhynchocephalians ("beak-heads") are members of the Lepidosauria along with the lizards and snakes, the latter two comprising the Squamata within lepidosaurs. The Rhynchocephalia includes *Sphenodon* and its many ancient relatives. These animals are generally lizard-like but have skulls that are more rigid and a row of upper teeth running along the length of the palate row. Their teeth are also usually more robust, may bear long cutting edges, and, unlike most lizards, tend to lack deep roots.

The tuatara is mainly restricted to islands (about 35 of them) off the coast of the two main islands of New Zealand, the main island populations having been killed off following human arrival. Individual tuatara may be up to 60+ cm (2 ft) long and weigh up to 1 kg (2.2 lbs) and live more than 100 years in some cases. They are primarily carnivorous but are known to eat a wide variety of food types including eggs and many insects. They reproduce slowly, with each female laying eggs only about every four years or so. Juveniles have the briefly expressed condition of having a third, parietal eye on the top of the skull, an organ that probably serves a general light-sensing function and helps regulate physiological rhythms before it eventually gets covered in regular skin scales.[2] The main eyes are relatively large. But if any reptile could be described as cute (other than some tiny lizards), even the adult tuatara would be it. It is a shame there are so few of them left.

Rhynchocephalian Paradise

The tuatara *Sphenodon* is the only genus of rhynchocephalian left today. But the Mesozoic was arguably a rhynchocephalian golden age, when a variety of forms with different teeth and body shapes existed, more than 40 species. The diversity in size and morphology just of the skulls of these animals was impressive. There were small gracile forms with lots of small teeth (e.g., little *Diphydontosaurus* and *Gephyrosaurus*) known from the Late Triassic of northern continents (e.g., UK, Italy, Germany, US). There were small robust forms with a smaller number of large bladelike teeth (e.g., *Clevosaurus*) from the Triassic and Jurassic of Asia, Europe, Africa, and North and South America. Relatively large, herbivorous *Sphenotitan*, from the Late Triassic of Argentina, was probably a basal relative of later eilenodontines. Long- and low-skulled *Pleurosaurus* occurs in the Jurassic of Europe. *Palaeopleurosaurus*, from the Early Jurassic of Germany, was an elongate, apparently aquatically adapted rhynchocephalian. *Ankylosphenodon* from the Early Cretaceous of Mexico was also apparently elongate and aquatic and demonstrated pachyostosis, thickening of the bones sometimes seen in aquatic animals (such as manatees), often to assist with ballast. *Zapatadon* lived during the Early Jurassic in Mexico and was rather tiny, with a skull only 11 mm (0.44 in) long. The Early Jurassic *Cynosphenodon* from Mexico is only known from a partial jaw but, at least with respect to these parts of the skeleton, appears to be surprisingly similar to the modern tuatara. The variation in tooth shape and arrangement among these forms suggests a largely insectivorous to omnivorous diet for many, although insectivorous and herbivorous specialists existed too.

In the Late Jurassic–age rocks of Germany one finds a number of taxa: *Kallimodon* (fig. 6.1A); *Homeosaurus* (fig. 6.1B); *Vadasaurus*; *Leptosaurus*; and the aforementioned, very elongate, aquatic, and probably fish-eating *Pleurosaurus*, which was also much larger than most of its relatives at up

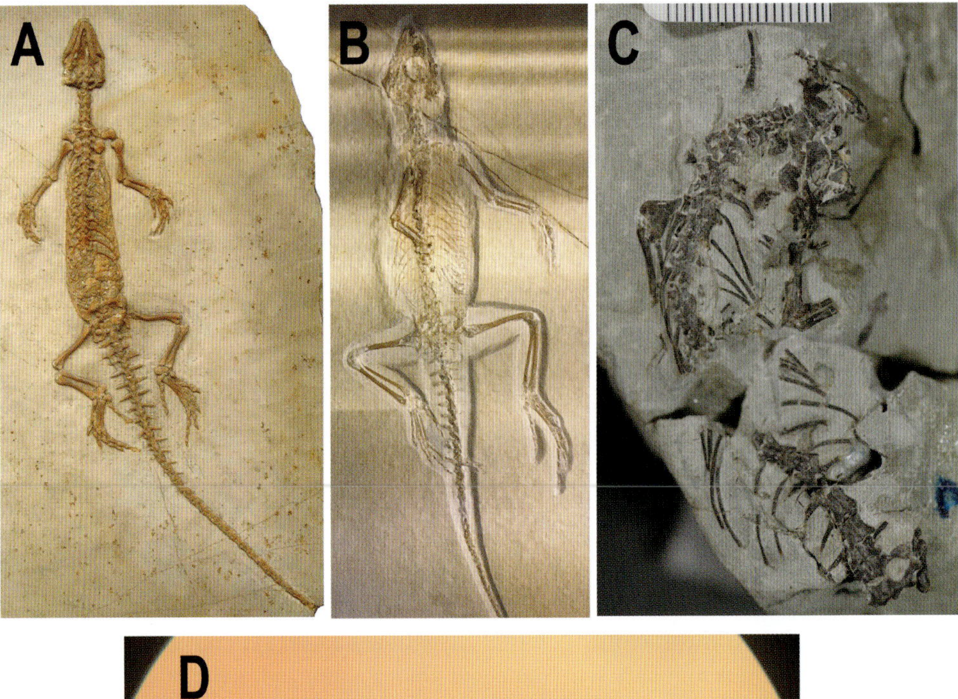

6.1 Some rhynchocephalians of the Late Jurassic. (*A*) *Kallimodon* from Germany (SNSB-BSPG 1887 VI 1). (*B*) *Homeosaurus* from Germany. (*C*) *Opisthiamimus* from the Morrison Formation of Wyoming (USNM PAL722041). (*D*) Dentary of *Opisthias* from the Morrison Formation of Dinosaur National Monument locality 375 in Utah. *Image in A courtesy of Marc Jones, C courtesy of Dave Demar, B and D by the author.*

to 1.5 m (5 ft) long. The *Sphenodon*-like taxon *Sphenofontis* was recently identified in Upper Jurassic deposits in Germany too, showing that there are still surprises in the rocks. Around the same time in France, the long-tailed, possibly toothless *Sapheosaurus* lived among the islands and tropical seas and reefs of Europe during that time when the Atlantic Ocean was only a few hundred miles wide and North America was so much closer.

6.2 Eilenodontine rhynchocephalians from the Upper Jurassic Morrison Formation. (*A*) *Eilenodon* specimen from western Colorado consisting of three cervical vertebrae, four dorsal vertebrae, a sacral vertebra and possible ilium fragment, two caudal vertebrae, and possible tibia fragments (specimen located at Museums of Western Colorado, MWC 9977). (*B*, *C*) Partial dentary of the large, herbivorous sphenodont *Eilenodon* from Dinosaur National Monument in labial view (*B*) and occlusal view (*C*). (*D*) Images of unworn tooth of *Eilenodon* from central Colorado (DMNH 10685) along with oblique views of partial dentary of same specimen. Color coding of *top-right* image shows enamel thickness of tooth; shading of dentary teeth drawing indicates top wear facet (purple) and side wear facet (blue). Scale in *A* = 5 cm, in *D* = 2 mm, and microscope view diameters in *B* and *C* = 15 mm; images in *D* from Jones et al., 2018, *Journal of the Royal Society Interface* 15: 20180039. Images in *A–C* by author.

Just across that ocean, back in the Morrison Formation, *Theretairus* was a rare rhynchocephalian. More common were *Opisthiamimus* (fig. 6.1C) and *Opisthias* (fig. 6.1*D*), both insectivorous to omnivorous forms, and *Eilenodon* (fig. 6.2), a large taxon with a robust skull and jaws and teeth with enamel of varying thickness along their length (fig. 6.2*D*), a feature that may have helped maintain the enamel-dentin shearing edge of the tooth as it wore down. *Eilenodon* was probably to a significant degree herbivorous. Although rhynchocephalians, at this point, appear to have been less diverse than lizards during Morrison times, they are overall more abundant as fossils and may have been more common within the ecosystem. The group is rare but present in Late Jurassic–age deposits in Portugal and Tanzania also.

Then there was *Oenosaurus*, yet another rhynchocephalian from the Late Jurassic of Germany, and one named recently by paleontologist Oliver Rauhut and others. The thing that makes this genus unique, among not just rhynchocephalians but all terrestrial vertebrates, is that the main upper and lower dentition consists of single, continuously growing tooth plates in each jaw, massive crushing surfaces reminiscent of the tooth plates of lungfish. These suggest that the animal was durophagus, feeding on very hard material of some type. Ancient aquatic taxa with presumed durophagus diets, such as lungfish and the Cretaceous shark *Ptychodus*, are generally thought to crush mollusk shells and crustaceans (freshwater or marine) as part of their feeding strategy, but what *Oenosaurus* was eating is less clear. With only skull material known and no postcranial bones, it is not even clear if it was terrestrial or aquatic. If it was terrestrial, it may have fed on snails and large insects. *Oenosaurus* appears to be related to mostly terrestrial taxa, although that doesn't necessarily mean it couldn't be aquatic.

By the Early Cretaceous, another large, possibly herbivorous eilenodontine, *Toxolophosaurus*, is found in the Kootenai Formation of Montana. Other Early Cretaceous occurrences include those in the Kirkwood Formation of South Africa and the Durlston of the UK, and two unnamed taxa, one apparently close to *Theretairus*, occur in the Early Cretaceous of Morocco as well. By the Late Cretaceous, the large, herbivorous eilenodontine *Priosphenodon* lived in what is now Argentina, and other Late Cretaceous forms from that country include *Kaikaifilusaurus*, *Patagosphenos*, *Lamarquesaurus*, and *Kawasphenodon*.

After the Cretaceous, rhynchocephalians obviously survived, as we still have them today, much to our good fortune. But their fossils almost completely drop off the radar after *Lamarquesaurus* and *Kawasphenodon* in the Cretaceous and Paleocene in Argentina. There are possible occurrences from the Paleocene of Morocco and West Africa, but other than that, the next occurrence is from 16- to 19-million-year-old Miocene deposits in New Zealand, and then Pleistocene fossils, again from New Zealand. And that's about it. So in comparison to the shrunken range, diversity, and abundance of the group since the Age of Dinosaurs, the Mesozoic was a veritable rhynchocephalian Xanadu.

Rhynchocephalians occurred on all continents except Australia and Antarctica, as far as we know, during the Mesozoic; their fossils will probably turn up in Antarctica. There were large and small forms, insect eaters, fish eaters, plant eaters, omnivores, and possible shell crushers; they lived on land and in the water. More so than with most other groups that survive today, rhynchocephalians truly demonstrate a former glory in diversity and disparity during the Mesozoic. But this only makes the modern tuatara that much more special to us.

Notes

1. This book's title is taken from the root meaning of *Theretairus* too, although unlike Simpson's intention with the genus name, here, I refer to dinosaurs rather than mammals as the "wild beasts."

2. Parietal eyes are not unique to tuataras. Many lizards, as well as some amphibians and fish, have these structures too.

Celebration of the Lizard (and Snake)

Squamata

7

RELATED TO THE RHYNCHOCEPHALIANS within the Lepidosauria is the order Squamata, consisting of the modern lizards, snakes, and amphisbaenians, all of which appeared alongside the dinosaurs at some point in the Mesozoic. For lizards, that was some time in the later part of the Triassic, although many crown-group lizards appeared by the Middle Jurassic. For snakes, that appearance was possibly by the Middle Jurassic, and for amphisbaenians, the earliest known are in the Late Cretaceous. The group evolved from their common ancestor with rhynchocephalians possibly by the Middle Triassic about 242 million years ago when we see the first definitive lepidosaur. In today's world, squamates rival perciform fish as the most diverse order of vertebrates.

Lizards and snakes are widely distributed today, of course. I grew up seeing western fence lizards (*Sceloporus occidentalis*) in woodpiles in our yard, gopher snakes (*Pituophis catenifer*) in the grass nearby, and sidewinders in the Mojave Desert (*Crotalus cerastes*). Being young, I assumed the gopher snakes were rattlers and initially ran like a madman without sticking around to find out the identification details. In the Rocky Mountains and Colorado Plateau, paleontological fieldwork to find ancient lizards and snakes often brings us into contact with their living descendants, and our field hosts in a way: collared lizards (*Crotaphytus collaris*), prairie and midget faded rattlers (*Crotalus viridis* and *C. concolor*), and plenty of others. We see lizards and snakes today under rocks, in brush and woodpiles, and on dry lakes. And these settings are in deserts, grasslands, orchards, and among redwoods—there are few environments that don't have squamates, the limitations to range mainly being extreme cold. This may in part be because there are few environments that don't have insects or plants, depending on the diet of the lizard species. Many snakes follow small mammals, of which there are often plenty. Some snakes have also invaded the water and become as mobile in ponds, swamps, and rivers as they are on land. And then there are the sea snakes of the Indian Ocean and Pacific, which are almost fully aquatic.

The more than 4,200 species of lizards today range from tiny forms up to the massive Komodo dragon (*Varanus komodoensis*) at 3 m (10 ft) long. The smallest modern lizard, *Sphaerodactylus ariasae*, is less than 2 cm (0.8 in) long. The lizards are represented in the fossil record by taxa that cannot always be classified into low-level groups such as families, especially among Jurassic species. Still, broad group assignments are often possible. Modern lizards are classified into several broad groups. Iguanians *may* be the most basal group, but there are significantly conflicting data in morphological versus molecular studies; the iguanians include agamids, chamaeleonids, and iguanids. Scleroglossa includes the geckos and Autarchoglossa, the latter of which is split into scincomorphs and anguimorphs. The scincomorphs include lacertoids (lacertids, teiids, and others) and scincoids. The anguimorphs include varanids and several other families. Snakes may have evolved from within the Anguimorpha, but available evidence is still a bit unclear on their precise origins. And current evidence suggests that amphisbaenians (chubby, legless burrowing squamates) may have evolved from the Scincomorpha.

Lizard fossils, especially in flood-based terrestrial deposits of the Late Jurassic through the Late Cretaceous of North America, are often just isolated elements with jaws, jaw fragments, and isolated vertebrae as the most easily attributable bones. Working in the Morrison Formation, we can find mostly isolated vertebrae, but then on rare occasions, we encounter an articulated partial skeleton—though only at certain sites. But in the lake to lagoonal deposits of what were calmer and more extensive ancient settings of China, Brazil, Germany, and a few other places, we occasionally see whole skeletons.

Lizard Kings

High above Salt Lake City in Utah, looking out over the metropolis from the lower slopes of the Wasatch Mountains east of downtown, is the Natural History Museum of Utah, sitting near an old shoreline deposit of the ice-age Lake Bonneville. Resting quietly in a white cabinet in the paleontological collections room of the museum is a nondescript rock from outside Hanksville, a tiny town north of Lake Powell that gets a significant amount of traffic from folks on their way to and from boating the desert's aquatic playground filling the former Glen Canyon. The rock is from the Morrison Formation, the layer that brought us *Brontosaurus* and *Stegosaurus*, among many others. But lurking in that rock when one looks closer is the small skeleton of a slender lizard, laid out as if it had desiccated in the rock last week—skull, vertebrae, ribs, and limbs, back to the base of the tail, all in position as they were when the animal scampered around the Jurassic landscape 150 million years ago (fig. 7.1*B*). It is one of very few lizards from the Morrison that is more than a jaw fragment or isolated vertebra (fig. 7.1C), and the record from the Cretaceous of North America is not all that much more complete. The specimen is thus a rare window to the ancient lizards that lived in the dinosaurs' world.

***Facing,* 7.1** Some lizards of the Mesozoic I. (*A*) The stem gecko *Eichstaettisaurus* (SNSB-BSPG 1937 I 1) from the Late Jurassic–age Solnhofen Formation of Germany. (*B*) An indeterminate lizard from the Upper Jurassic Morrison Formation of southern Utah (UMNH VP13829). (*C*) Indeterminate lizard legs and tail from the Morrison Formation, Fruita Paleontological Area, western Colorado. (*D*) The Early Cretaceous *Scandensia* (MCCM LH-11001) from Las Hoyas, Spain. (*E*) Possible gekkotan lizard skull from the Morrison Formation of Dinosaur National Monument (DINO 15914). Scale in *A* in centimeters; scale bar in *B* = 1 cm. *Photo in A courtesy of Randy Nydam; photos in B and E by author, courtesy of the Natural History Museum of Utah, Bureau of Land Management, and Dinosaur National Monument; photo in C courtesy of George Callison; photo in D courtesy of Susan Evans.*

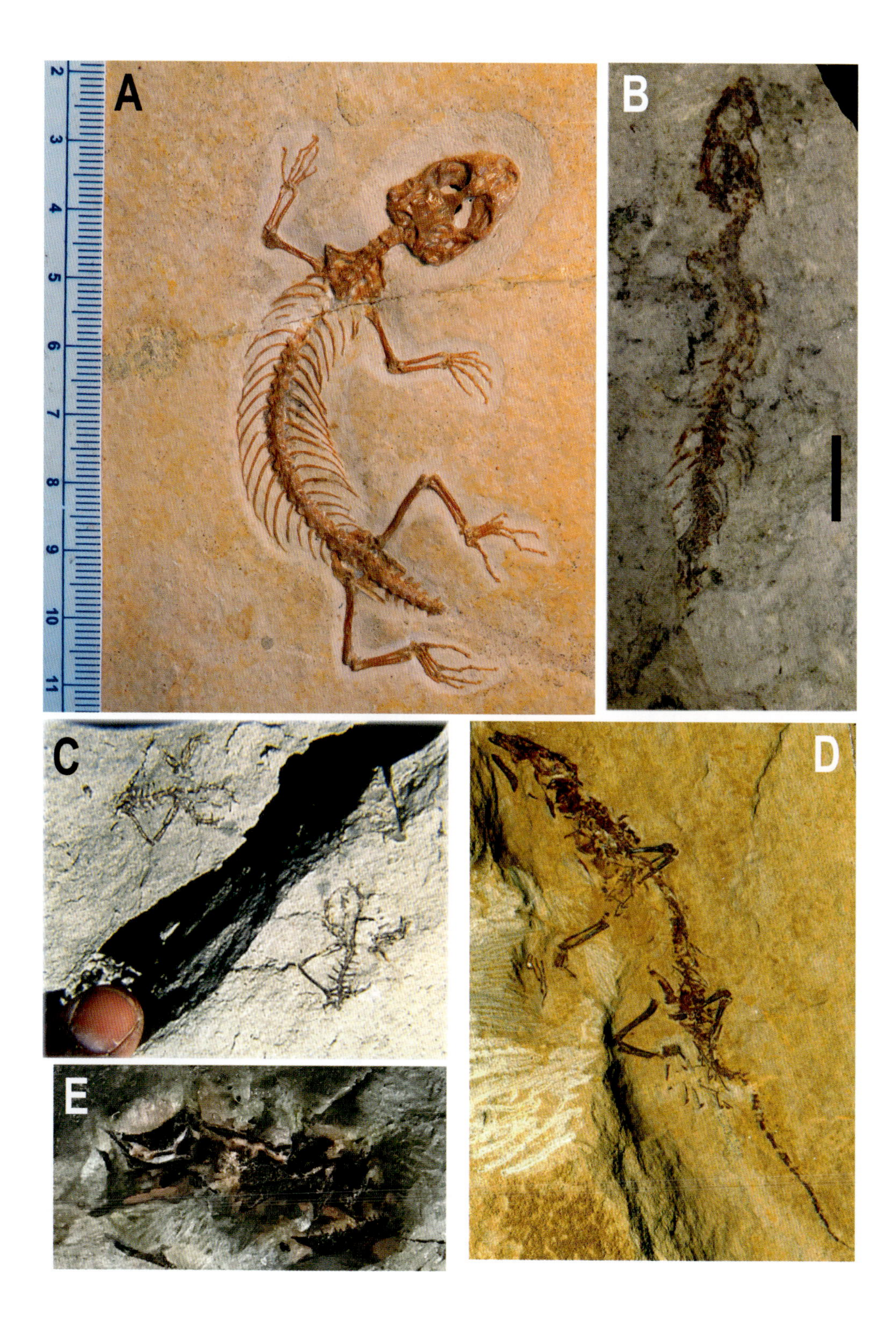
A
B
C
D
E

While lizard fossils are not abundant in most Mesozoic formations, it is likely because they are small and often destroyed before fossilization, and once exposed, they are hard to find due to their small size—it is almost certainly not because they were uncommon or not diverse at the times and places they are found. Snakes were probably limited in diversity and somewhat rare, however. In 30 field seasons working in Mesozoic formations of western North America, I can count on my fingers the number of lizard fossils I've found—a jaw in the Jurassic of Wyoming; a handful of individual, isolated vertebrae in the Fruita Paleo Area in Colorado; and maybe one or two from Late Cretaceous formations. They aren't exactly falling out of the outcrop in most of North America or anywhere else for that matter, and few specimens are nearly complete.

In the early 2000s, northern Italy's Middle Triassic rocks produced a lizard-like stem squamate in *Megachirella*, an animal that appears to be ancestral to later lizards, snakes, and amphisbaenians. So squamates, like rhynchocephalians, turtles, and lissamphibians, appear to have arisen around the same time as dinosaurs. *Megachirella* is known from a single specimen consisting of the front half of the animal. It appears to have been terrestrial and had fairly robust forelimbs and a relatively large triangular skull.

The first hints of true lizards may be coming from the Chinle Formation of the Late Triassic of North America, but these have not yet been published. The fact that rhynchocephalians have been found as far back as the Late Triassic suggests we should be able to find at least stem lizards from this time, but so far, the oldest true lizard fossils described date from the Middle Jurassic. We might reasonably expect Late Triassic lizards to be found or officially reported soon.

The Middle Jurassic starts ushering in some recognizable true lizard groups. In England, the Middle Jurassic Forest Marble Formation has produced the scincomorphs *Paramacellodus* and *Saurillodon*, not the last we will see of those genera but the earliest true lizards currently known. That these first lizards are scincomorphs and anguimorphs and not the more primitive (earlier branching) iguanians or gekkotans suggests that the latter, less derived two groups still elude us in older rocks. In Scotland, the Kilmaluag Formation has yielded the scincomorphs *Balnealacerta* and *Bellairsia*, known mainly from jaw fragments. In China, *Hongshanxi* is a stem scleroglossan lizard known from a nearly complete articulated skeleton; it was probably a predator of larger insects and lived on and near the ground, perhaps frequently scrambling around rocks or bushes as well.

In the Late Jurassic–age Morrison Formation of the western US, in addition to the unidentified slender lizard at the Natural History Museum of Utah, there are both scincomorphs (*Paramacellodus*, *Saurillodon*) and anguimorphs (*Dorsetisaurus*), along with one indeterminate form (*Schillerosaurus*) and a possible gekkotan, represented by a partial skull (fig. 7.1*E*). *Paramacellodus* is represented by material from Wyoming and Colorado. Rocks of the same age in Portugal also preserve *Dorsetisaurus* and *Saurillodon*, plus the scincomorph *Becklesius*. But most of these are the isolated jaw or skull preservation I mentioned above; a few nearly complete lizards from the Morrison are still mostly unidentified. To see full lizard morphology from 150 million years ago, we need to go to the Late Jurassic–age units in Germany, where there are still relatively few specimens in about five genera, but where they are buried in shallow lagoon deposits and are more complete. Little *Eichstaettisaurus* (fig. 7.1A), a relative of the geckos, shows that ancient Dinosaur Age lizards were quite similar in size and shape to modern lizards, although they may have been different in a few morphological details here and there. Some of these lizards may have already developed gecko-like gripping characteristics of the hands and feet.

At about 11 cm (4.5 in) long, the type specimen of *Eichstaettisaurus schroederi* is interesting in that only the first five tail vertebrae are preserved, despite the rest of the skeleton being complete, and there appears to be a "shadow" of the rest of the tail in the limestone slab—it may have lost part of its tail and regrown it as soft tissue only. Lizards today, of course, can lose parts of their tails and regenerate them (autotomy). This ability results from the development of special adaptations of the tail vertebrae and soft tissue, adaptations that fossil evidence demonstrates appeared in lizards at least as far back as the Eocene. And it appears to have been present in at least one basal diapsid (relative of lizards) during the Early Jurassic. So, some Mesozoic lizards may well have developed this ability as well. Although it is preserved in shallow marine rocks, *Eichstaettisaurus* appears to have adaptations for a combination of climbing trees and bushes and living on the ground (scansorial, as opposed to fully arboreal).

The Late Jurassic epoch saw a relatively uniform lizard fauna extending from North America, across Europe, and into Asia. Many of the same groups, and in some cases genera, are known from multiple regions. This persisted through the Early Cretaceous when the splitting of the supercontinent Pangea was sufficiently complete to isolate the respective continents.

By the Early Cretaceous, the lizards of the UK's Purbeck Group included some of the same forms of the second half of the Jurassic: *Becklesius*, *Dorsetisaurus*, and *Paramacellodus*. Elsewhere, more records have started showing up in the Early Cretaceous too. Three lizard genera are well preserved in the lake deposits of the Santana Group in Brazil, one a probable scincomorph and the other two as yet enigmatic. *Scandensia* (fig. 7.1D) is known from the Las Hoyas locality in Spain. The Kem Kem Group of Morocco recently yielded an acrodontan iguanian, *Jeddaherdan*, a form that, as an iguanian, is strikingly more basal among lizards than most of the geologically older lizards we've encountered so far in this book. In the Cedar Mountain Formation of Utah on the Colorado Plateau, lizards include seven species of mostly indeterminate teiids, scincids, paramacellodids, and others.

The Early Cretaceous–age Jehol Group of China has five species of lizards in its lake deposits. Among these is *Xianglong*, an iguanian that appears to have been a gliding lizard. Preserved as a nearly complete, articulated skeleton, *Xianglong* had extra-elongate ribs that it presumably could splay out, with a skin-gliding membrane between to swoop from tree to tree or to the ground. The longest of these ribs are longer than the forelimbs and

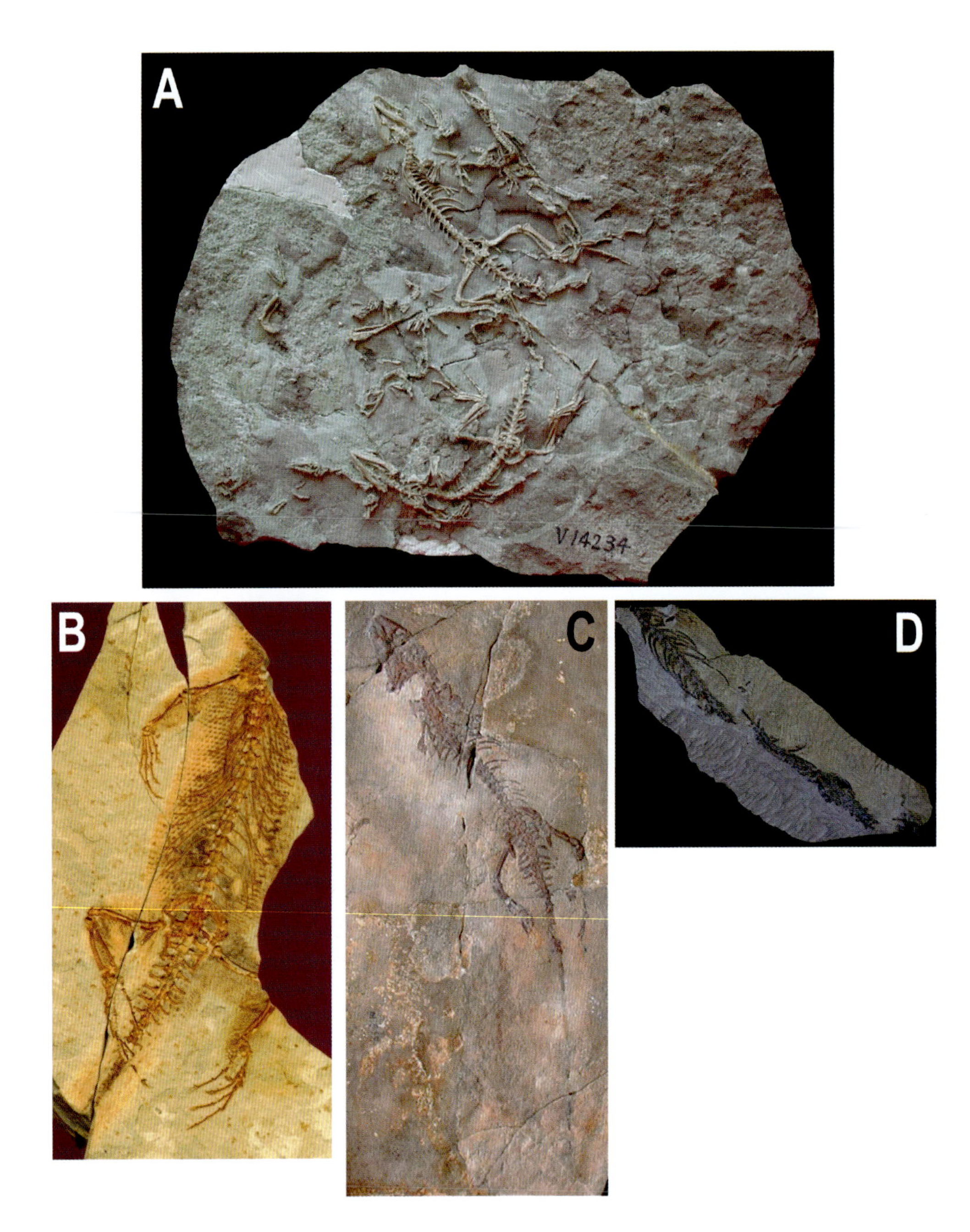

7.2 Some lizards of the Mesozoic II. (*A*) Multiple individuals of the anguimorph *Dalinghosaurus* (IVPP V14234) from the Lower Cretaceous Lujiatun Formation of China. (*B*) *Liushusaurus* (IVPP V14715B) from the Early Cretaceous Jehol Biota of China. (*C*) The anguimorph *Chometokadmon* (MPN 539) from Pietraroia, Italy (Early Cretaceous). (*D*) The aquatic *Kaganaias* (SBEI 12459) from the Lower Cretaceous Kuwajima Formation of Japan. *All images courtesy of Susan Evans; D © Shiramine Institute of Paleontology.*

about as long as its hind limbs. The skin flap, or patagium, appears to have stretched between the bases of the forelimbs and hind limbs and to have been extended out by rib movement when needed, like a swing-wing jet. The ribs seem to have swung somewhat back to front when deployed, and the movement apparently was not ventral to dorsal. The patagium also appears to have been supported between bones by collagen fibers.

The Jehol also preserves the anguimorph *Dalinghosaurus*, a form with hind limbs that are unusually long relative to the forelimbs (fig. 7.2A). This may have been an adaptation to climbing or possibly to being a ground-dwelling runner. Recent analyses of *Dalinghosaurus* have also determined that it is likely closely related to the knob-scaled lizards (xenosaurids) among anguimorphs. Members of this genus have even been found preserved in an aggregation of parts of 16 individuals of a range of ages. This slab may be a chance taphonomic concentration of the skeletons, but if not, it may well represent members of the lizard species sheltering to try to escape a volcanic eruption. We may be seeing evidence of 125-million-year-old behavior in a lizard species with this specimen preserved at Beijing's Institute of Vertebrate Paleontology and Paleoanthropology. Another specimen from China, Late Jurassic to Early Cretaceous in age and found in a unit with numerous salamander specimens, almost perfectly preserves the outline of the lizard's body in skin impression. Another Jehol taxon, *Liushusaurus* (fig. 7.2*B*), preserves traces of the skin and scales of the body.

Other lizards known from around the world, among many, include anguimorphs from the Early Cretaceous of Italy (fig. 7.2C) and an aquatic form from the same epoch in Japan (fig. 7.2*D*). These are just a few of the better-preserved fossils. There are many others.

As the Late Cretaceous proceeded, lizards diversified, and many specimens clearly representing modern groups appeared. A good example of this is the Campanian-age lizard fauna of the Djadokhta and Barun Goyot Formations of Mongolia. Here, paleontologists have encountered more than 25 species of iguanid, acrodontan, gekkotan, teiid, scincoid, scincomorph, anguimorph, and varanid lizards, far more diversity than we saw in formations previously—and closer to today's forms. Many of these Gobi Desert specimens are spectacularly well preserved and are sometimes nearly complete, often articulated, and consistently three-dimensionally preserved in sandstone, still nearly as white as the bones were in life. Elsewhere, Late Cretaceous lizards have been reported from Argentina, Madagascar, and Hungary, and from amber deposits in Asia. Dolichosaurids are aquatic lizards close to varanoids, mosasaurs, or snakes and are known from Late Cretaceous deposits in Europe and Lebanon. The rise of the "Modern Fauna" of lizards showed similar themes in North America and Asia even though the taxa are very distinct. Obviously, some occasional biological communication between North America and Asia occurred in the Late Cretaceous.

My introductory summer of fieldwork with Dave Archibald back in the day involved some quarrying on a hadrosaur dinosaur, but a lot of our work was hauling out and screenwashing anthills, mainly in search of Cretaceous mammals. We only saw one tooth all summer, however, the mammals remained otherwise elusive, only revealing themselves after additional washing and picking back in San Diego. It turned out that we had also found a couple of lizard jaw fragments such as *Leptochamops* (our crew or others in previous or subsequent years, at least), but again, we never saw them in the field. And we have recently found at least one tooth of a teiid lizard in screenwashing work at a new site in the same area. The Williams Fork Formation is the same age as the Djadokhta and Barun Goyot but so far has produced far fewer lizards and much less diversity, but we're working on that. Certainly, the environments were different. In northwestern Colorado, the Williams Fork sediments were laid down by river channels and coal swamps of a major river delta prograding out into the Western Interior Seaway—think southern Louisiana today. In what is now the Gobi Desert of Asia, the environment consisted of rivers and floodplains mixed with abundant sand dunes in an arid to semiarid environment that was perhaps similar to the oasis and surrounding desert of the Okavango Delta of Botswana.

Plenty of lizards equally as fragmentary as those out of the Williams Fork have been found in other Late Cretaceous formations of the Rocky Mountain region. The Dakota, Wahweap, Kaiparowits, Aguja, Dinosaur Park, Wapiti, Mesaverde, Hell Creek, and Lance Formations all frequently produce lizard vertebrae, jaws, and teeth from screenwashed microvertebrate sites. Many of these are scincomorphs and anguimorphs, and the latest Cretaceous (Maastrichtian) in North America alone has more than 25 genera of iguanids, polyglyphanodontians, and scincomorphs and anguimorphs. The octosyllabic polyglyphanodontian lizards stood out among their peers in evolving complex, almost mammal-like cusped teeth before becoming extinct at the end of the Cretaceous. Among the anguimorphs was *Palaeosaniwa*, an ancient varanid lizard that was related to the Komodo dragon and could grow to the same intimidating sizes. Additionally, a relative of the Gila monster has been reported from the Late Cretaceous of Utah.

Our lizard friends today exhibit a wide range of ecologies. Most reproduce by laying eggs, but up to 20% of all lizard species nowadays are viviparous and give birth to live young. Members of one South American and Caribbean genus (*Mabuya*) have even developed a full-formed placenta for their embryos! Almost all lizard species reproduce sexually, but a good baker's dozen of genera (including 30 species total, among them the New Mexico whiptail, *Cnemidophorus neomexicanus*) are parthenogenetic—they reproduce as all-female, clone species without males in the population. Five of these 13 genera are geckos. This results from the development of an embryo from an unfertilized egg, and it may occur in a previously sexually reproducing species, the males simply disappearing from the population as more descendant females begin cloning themselves too. It sounds like a science fiction movie (*Attack of the Clone Lizards?*), but it happens! Parthenogenetic lizards are able to increase population sizes at an exponential rate far outstripping closely related, sexually reproducing species. The tradeoff is that parthenogenetic species are more susceptible to extinction as their environments change because they have far less genetic diversity within the population. So these types of lizards may be short-lived as species, or they may last for significant periods of time in stable environments.

While many lizards are opportunistic predators of insects and other taxa, there are specialized groups. Some tropical iguanas feed mostly on plants. Some lizards will sometimes eat other lizards, including their own kind, presumably on very rare occasions. Some species eat more than 50% ants, others eat almost no ants at all. Some eat spiders and scorpions, while one snakelike pygopodid species eats only other lizards. One study of the stomach contents of 445 individuals of a species of teiid lizard from the Neotropics found among the 4,478 total items (about ten items per individual stomach) a wide variety of insects as well as spiders, mites, scorpions, centipedes and millipedes, isopods, mollusks, annelids (earthworms), and plant material. The primary diet items were termites (34%), insect larvae (16%), and grasshoppers and crickets (9%). This largely insectivorous and predatory species still consumed about 1.25% plant material.

Other specializations developed by various groups of modern lizards include herbivory, venom injection, telescopic eyes and a ballistic tongue (chameleons), body armor, aquatic habits, arboreality, gliding, fossoriality, nocturnality, and leglessness, among others. Some of these may have been present in Mesozoic species.

Why Does the Floor . . . Move?

More than a few people around the world are a bit edgy around snakes, from (reportedly) Matt Damon to the fictional Indiana Jones ("I hate snakes, Jock. I hate 'em!") to, well, me. The talent to move along, sometimes at a rapid clip, without limbs of any kind is an ability some consider to be unnatural—or at least that's my personal reason that I get a little wigged out around them. The ability to rock climb without limbs seems supernatural also.[1] The fact that some are venomous probably doesn't help anyone. Of course, moving *sans* limbs is entirely natural, as leglessness has evolved multiple times, in the varanoid and pygopodid lizards (snakes and amphisbaenians) and in amphibians (caecilians). Whether ophidiophobia is innate or learned is not clear. Some studies have found that between one-quarter and one-third of people around the globe are afraid of snakes, so it might make sense that we have simply evolved with a natural leeriness about them due to the obvious danger posed by venomous species. But myths, legends, folklore, and bad press about reptiles in general have probably contributed as well, and some studies have found confirmation of this. Those of us not entirely at ease around snakes may have both evolutionary and cultural reasons for our wariness. Regardless, the rise of fossil snakes is a fascinating story.

Snakes evolved from an unknown group of lizards, possibly from among the varanoid anguimorphs, sometime in the Middle Jurassic or possibly the Early Cretaceous. The elongation of the snake body and the eventual loss of the limbs may have been adaptations for an initially burrowing lifestyle, but snakes may instead have evolved in an aquatic setting. There is some debate on this, and the issue doesn't seem to be entirely resolved just yet. There are more than 3500 species of snakes today; many hunt small mammals and other tetrapods and swallow them whole. The lower jaws are connected to each other at the front by a very loose ligament that allows the snake to expand its mouth and get the jaws around very large prey items. This allows snakes to swallow prey larger than their skulls and "walk" it down the throat with their skull and lower jaws. There are venomous species, constrictors, and more benign forms, though all snakes are carnivorous. They are characterized by an increased number of vertebrae with interlocking additional zygapophyses, an expandable pair of lower jaws, sharp conical and strongly recurved teeth, and a complete loss of forelimbs along with near-complete loss of the hind limbs. Most of these characteristics are also true of Mesozoic snakes, but more complete specimens indicate that limb loss was one of the later attributes achieved by snakes. Their bodies got very long, and they began to slither, but a number of Mesozoic forms had not yet lost their limbs entirely; they just retained shrunken versions of them.

Tetrapodophis from the Early Cretaceous of northeastern Brazil is a nearly complete fossil with almost 150 vertebrae, and it retained tiny forelimbs and hind limbs. It was originally identified as a stem snake but appears now to be a dolichosaurid, a member of a group of small, aquatic varanoid lizards closely related to mosasaurs or snakes. The snake *Eupodophis* is also articulated and nearly complete and was found in the Late Cretaceous–age Sannine Formation in Lebanon; it retains reduced hind limbs. *Najash* from the Late Cretaceous of Argentina had some of the least vestigial limbs of all ancient snakes found so far, retaining the sacrum and with bone extending beyond the rib cage. Retention of limbs in at least some snakes seemed to span the group's Mesozoic history.

In addition to reduced hind limbs, another interesting aspect of the Lebanese Cretaceous snake *Eupodophis* is that it was a sea snake, with pachyostotic (thickened) bones for ballast and a laterally compressed tail for propulsion. Other Cretaceous sea snakes, *Pachyrhachis* and *Haasiophis*, are known from not far away in the West Bank. Both possess reduced hind limbs. All three of these sea snakes would have occupied the coastal seas of the Tethys Ocean about 95 million years ago during the Late Cretaceous. It is not clear how closely related these snakes are to modern groups—or, at least, there is not yet much of a consensus on whether they are stem or crown taxa—and so we are not sure if their aquatic habits imply that the origins of snakes were also aquatic.

Most of these snakes are about 30 cm (12 in) long, give or take. Nothing too tiny or giant. But some Late Cretaceous snakes appear to have gotten quite large. Vertebrae of *Madtsoia* from India are of a size suggesting that the full animal may have been 5 m (16 ft) long. Other specimens of

Madtsoia have been found in the Late Cretaceous of Niger, Madagascar, and Spain.

Coniophis was named from the Late Cretaceous–age Lance Formation in Wyoming by O. C. Marsh in 1892. Not much of its structure is known because although it has been reported from many areas, its remains tend to be isolated occurrences, and no articulated or particularly complete specimens have been found. Since Marsh's description, it has been reported from the Late Cretaceous of Canada, Montana, New Mexico, and Utah in North America. Other reported Cretaceous occurrences include India and Sudan. But *Coniophis* has also been reported from Lower Cretaceous rocks in the Cedar Mountain Formation of Utah. A recent description of more complete material indicates that *Coniophis* was a primitive, burrowing snake with a long body and an only slightly flexible lower jaw; this suggests that body elongation preceded the fully mobile snake jaws.

Early Cretaceous–age deposits are not necessarily the beginning of the story for snake origins, it seems. Recently named snakes seem to push their appearance back a bit. The Late Jurassic of Portugal has *Portugalophis*, and the contemporaneous Morrison Formation in North America now has *Diablophis*, both believed to be snakes with vestigial limbs based on jaws and vertebrae found in the deposits. The Late Jurassic to Early Cretaceous–age Purbeck Formation of southern England has *Parviraptor*, possibly a similar snake. And the Forest Marble Formation of Kirtlington, England, has *Eophis*, an apparent Middle Jurassic snake. But these are not without controversy, and there are some squamate paleontologists, and not necessarily a small minority, who suggest that these are not snakes at all but are simply possible anguimorph lizards with highly recurved, conical teeth similar to snakes. The dust has yet to settle, metaphorically speaking.

Squamate Trends

The lizards and snakes demonstrate some trends common in ancient paleobiology. In the case of lizards, those would be: 1) an apparent origin around the Triassic–Jurassic boundary or earlier, common for many of the microvertebrate groups featured in these chapters and 2) that deep origin being reflected in seemingly missing fossils—representatives of the most primitive lizard groups are not necessarily showing up as the geologically oldest in many faunas. There should be more to find in older deposits. Lizard faunas also seem to become more cosmopolitan through the Cretaceous. And new discoveries are likely in southern continents where relatively fewer taxa have been reported thus far.

As for snakes, they show a mild degree of mosaic evolution, acquiring traits at different times in a transitional process, for example, rather than smoothly, gradually, and all together. This might be described as evolution by fits and starts. In the case of the snakes, they seem to have lengthened their bodies and increased their total number of vertebrae before fully reducing and then losing their limbs. In addition to elongating their bodies first, they developed greatly recurved and conical teeth a bit before developing their mobile skull and jaw joints.

The squamates are respectably diverse as fossils, with gliders, large carnivores, and stem taxa of many modern groups present. It is probably safe to infer that they were even more splendidly diverse as living animals—we are likely only seeing a small percentage of those species that lived during the Mesozoic.

Note

1. I once witnessed a hunting snake moving into and out of individual swallow nests lined up like a village and tucked well up under a sandstone overhang right above the Colorado River. I have no idea how the snake got down to this sheltered alcove directly above the water, but it was an impressive feat of snake bouldering.

In the Realm of Poseidons

Marine Reptiles

8

IF ANY SETTING OF THE MESOZOIC matched that of the dinosaurs for toothy danger, it was the seas. Just when you thought it was safe to go back in the water. It was far worse than just the giant fish *Xiphactinus* and some sharks. Whereas many of the groups we've visited so far reflect the familiarity of the faunas (aside from the dinosaurs) on land, the large animals that lived in the seas of the time matched the dinosaurs for sleek, exotic design; predatory danger level; and the inexplicable (to some of us) human proclivity for broad herpetophobia. The sea monsters of the Age of Dinosaurs were not dinosaurs, but they were big, reptilian predators—some of which effectively slithered through the water—that have many of the same psychological effects on us that their dinosaurian cousins can have. But the marine reptiles that ruled the oceans of the Mesozoic were not sea monsters. They were quite real, and they played ecological roles that were to some degree new and that were later filled by animals such as dolphins, orcas, sperm whales, and possibly gray whales.

Fish Lizards

The Mesozoic was a time when the seas of Earth featured reptilian "dolphins" known as ichthyosaurs ("fish-lizards"). First illustrated from finds in Europe in the late seventeenth century, but not recognized as a unique reptile until the early nineteenth, the ichthyosaurs had a global distribution in the seas during most of the Mesozoic and helped make a 12-year-old British girl a household name among fossil enthusiasts, young and old alike, a full 200 years after she was putting marine reptiles and other fossils on the map. But the ichthyosaurs were a respectably diverse group of creatures, and they had a sleek beauty that matched that of Flipper, Winter, Willy,[1] and other famed representatives of their ecological understudies of the modern mammalian world among the dolphins and toothed whales.

The prototypical ichthyosaur would be *Ichthyosaurus*, a reptile 3 m (10 ft) long, with a body shape not unlike a modern dolphin: narrow snout with many conical teeth, pectoral and dorsal fins, reduced pelvic fins, short body, and sharklike double-lobed caudal fin for propulsion (only with the vertebral column curving down into the ventral lobe instead of the dorsal lobe, as in primitive fish). The tail moved side to side with that vertical tail, however, unlike the horizontal tail and up-and-down motion of the dolphin tail; this is a difference born of differences in the movements of the spinal columns of reptiles versus mammals. The eyes were comparatively larger than those of dolphins or sharks, however. The small ichthyosaurs were reptiles that, like modern dolphins, were probably fast-swimming predators of fish and squid, primarily. This body shape appears to have developed through the phylogenetic development of the group—basal taxa tend to be more elongate in body and tail, and the most dolphin-like shapes are achieved in more derived ichthyosaur taxa.

The first complete ichthyosaurs, the first ones that indicated that they were neither fish nor a known type of reptile, were reported from Europe and the UK in the early nineteenth century (fig. 8.1*A*). Along the southern coast of England, some of the best specimens were found in the Lower Jurassic by Mary Anning of the town of Lyme Regis. Anning was from a family who supplemented their income by finding and selling small fossils from the cliffs along the sea, but she took it to a new level with her discoveries of larger and scientifically significant specimens. Her first ichthyosaur, found when she was 12 and excavated by herself and a crew hired by the family, was nearly 5 m (16 ft) long and turned out to be the body that went with a skull that she and her brother had found a bit earlier. She collected specimens from this area for most of her life and ended up finding multiple articulated ichthyosaurs, as well as plesiosaurs, pterosaurs, fish, and invertebrates.

Ichthyosaurs appeared in the Early Triassic in Europe, Asia, and North America and were more widely distributed in the same continents (plus New Zealand) by the Middle to Late Triassic. Ancestral protoichthyosaurs named *Utatsusaurus* are known from the Early Triassic of Japan (fig. 8.2*A*), and these dolphin-sized animals are large for the early forms, many being dog sized. Full ichthyosaurs were common through the Jurassic in Europe and also appeared in South America during that period. They also occur in the Jurassic in western Asia and North America, with a more limited number of genera ranging into the Early Cretaceous but on all continents except Africa and Antarctica. They appear to have become extinct sometime in the early Late Cretaceous during the Cenomanian age about 100 million years ago, one of the last forms being the widely distributed genus *Platypterygius*.

Throughout this time, there were many forms similar to *Ichthyosaurus* in general body form (fig. 8.1*C* and 8.1*D*): from *Stenopterygius* in the Early to Middle Jurassic of Europe (fig. 8.1*E*), a form that demonstrates that baby ichthyosaurs were born live, tail first, into their oceanic environment; to seven species of the nearly 7 m long (23 ft) *Platypterygius* from the Early to Late Cretaceous, a form that fed on birds and sea turtles, presumably among other taxa. In addition, there are a number of other forms from the Triassic through the Early Cretaceous in Europe, Asia, and North and South America. Among these is the 5 m (16 ft) *Ophthalmosaurus* from the Middle Jurassic, also of Europe and North and South America. In North America, *Ichthyosaurus* is known from the Early Jurassic of Alberta, and *Ophthalmosaurus* has been found in the Middle to Late Jurassic Sundance and Stump Formations of Wyoming and Utah. These were the "dolphins" of the Mesozoic seas, but there was more variety among the broader group. And a few of those were the largest reptilian sea monsters ever.

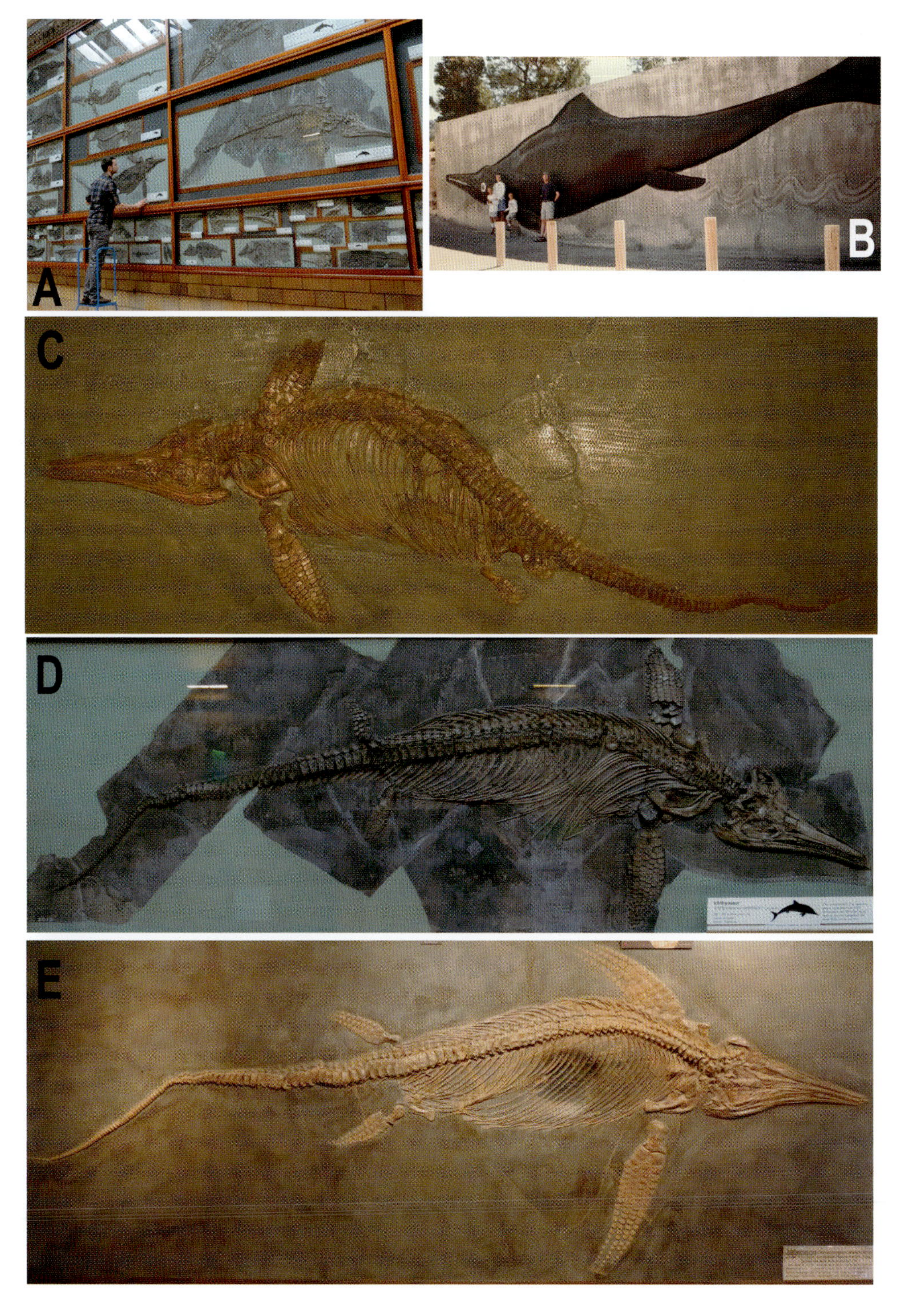

8.1 Some ichthyosaurs of the Mesozoic. (*A*) Marine reptile specimen wall at the Natural History Museum in London, paleontologist Dean Lomax for scale. (*B*) Life-size reconstruction of the Triassic ichthyosaur *Shonisaurus* at Berlin-Ichthyosaur State Park in Nevada. (*C*) Early Jurassic *Ichthyosaurus* from the Academy of Natural Sciences in Philadelphia. (*D*) Early Jurassic *Ichthyosaurus* from the Natural History Museum in London. (*E*) Early Jurassic *Stenopterygius* from Holzmaden, Germany, State Museum of Natural History, Stuttgart. *Photo in B courtesy of Chris Alexander; photos A and C–E courtesy of Dean Lomax.*

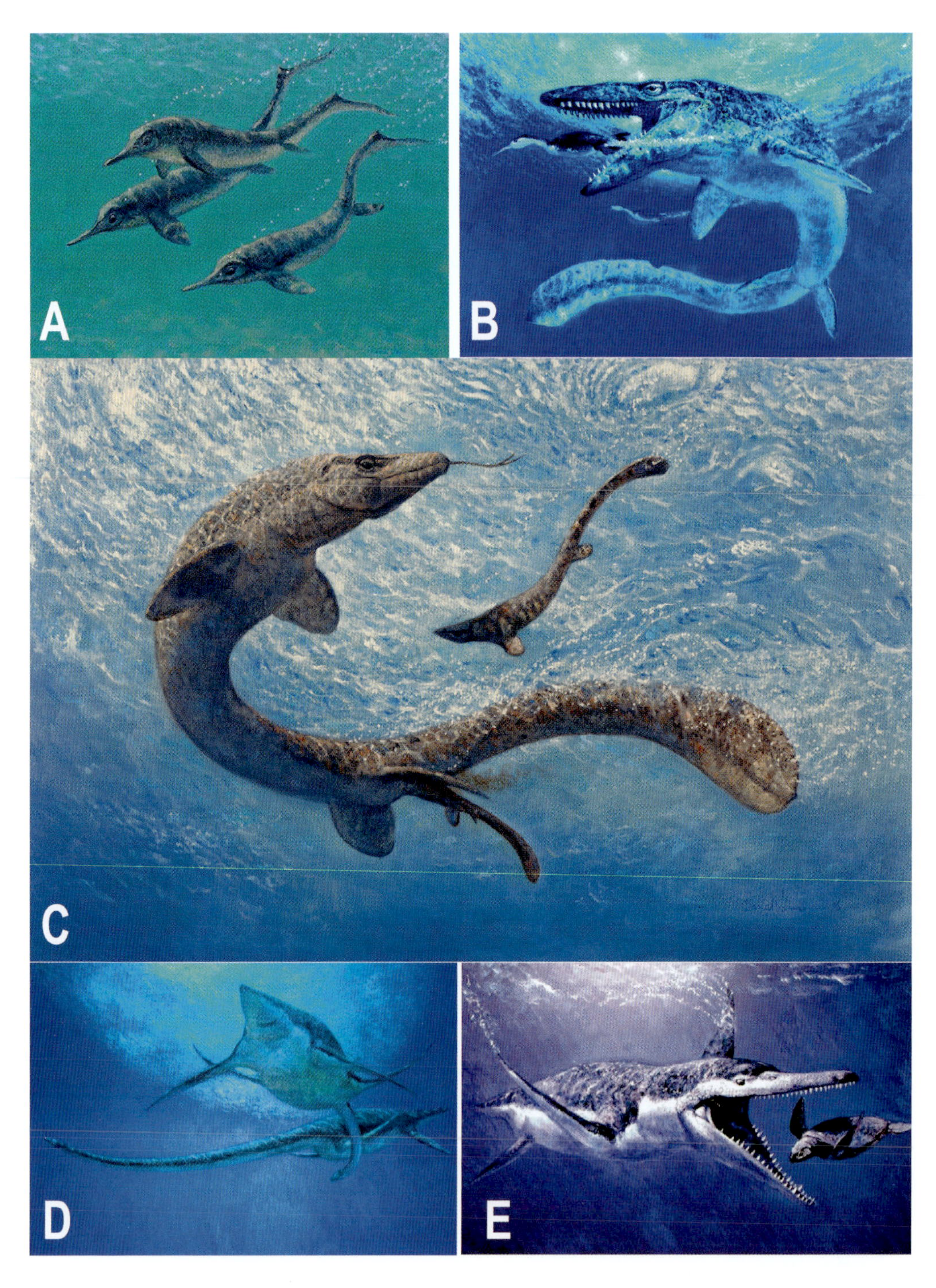

8.2 Marine reptile life reconstructions. (*A*) A pod of the Early Triassic ichthyopterygian (basal protoichthyosaur) *Utatsusaurus* from Japan. (*B*) The Late Cretaceous mosasaur *Tylosaurus* chasing the swimming bird *Hesperornis*. (*C*) The mosasaur *Plioplatecarpus* giving birth to the second of her baby mosasaurs. (*D*) The long-necked, elasmosaurid plesiosaurs *Styxosaurus*. (*E*) The short-necked, pliosaurid plesiosaur *Brachauchenius* lunging for a sea turtle. *All paintings by Dan Varner and courtesy of Mike Everhart.*

Somewhere around 130 mi (210 km) southeast of Reno, Nevada, in the Basin and Range province so well described by author John McPhee, a sentinel of a bygone mining age lies on a west-facing hillside scattered with the scrubby conifers typical of so much of the more arid parts of western United States: the former town of Berlin. About as much the opposite of its Teutonic namesake in Europe in every way imaginable, Berlin, Nevada, lies alone on the edge of the Shoshone Range above an inactive gold and silver mine. The town never housed more than about 300 residents and existed for only about 15 years, being abandoned by 1911. It survives as a ghost town today, part of a state park dedicated to it and a sea monster.

About a mile uphill from the ghost town of Berlin, paleontologists Charles Camp and Sam Welles from the University of California excavated the remains of almost 40 individuals of a giant fish-lizard, an ichthyosaur, named *Shonisaurus popularis* starting in 1954. These were found in the early Late Triassic–age Luning Formation, which represents a shallow sea on the edge of the ancient Pacific Ocean, much of the coastal sides of California and Oregon having been "slapped" onto North America since the Triassic thanks to plate tectonic action. The ichthyosaur specimens of the Shoshone Range had been found in the 1920s, lying in the hillside southeast of the mining town, and today, the site and some remaining specimens are protected from the elements by a building open to visitation. The dirt road into the site, though fine, can still eat tires now and then, but what lies in the building is quite impressive—strings of thin, spool-shaped vertebrae laid out in line in the rock like any good articulated skeleton, except that, in this case, the vertebrae are dinner plate–sized behemoths the size of sauropod dinosaur vertebrae. No reptilian dolphin in scale, *Shonisaurus* from Nevada was a whale-sized giant of some 15 m (50 ft) total length, with a skull alone nearly 3 m (10 ft) long (fig. 8.1*B*). The related *Shastasaurus pacificus* had been found previously in northern California, and another Late Triassic species of shastasaurid was collected from northern British Columbia, described by Elizabeth Nicholls and Makoto Manabe in 2004. It was even larger at up to 21 m (69 ft). Other related ichthyosaurs, including a third species of *Shastasaurus* have been found in China and the UK, and some of these suggest lengths up to 26 m (85 ft)!

Some shastasaurid ichthyosaurs appear to have been toothless as adults and possibly as juveniles. Recent finds in Nevada indicate that *Shonisaurus* had teeth at all ontogenetic stages, but the 10-m-long (33 ft) *Guanlingsaurus* appears to have been toothless throughout its life. *Guanlingsaurus* may have been a suction- or ram-feeding predator of squid, fish, or other prey, ecologically similar to modern sperm whales, narwhals, Risso's dolphins, and pygmy sperm whales, or many sharks. Toothed predators like *Shonisaurus* were probably traditional predators of relatively large prey. Unlike smaller ichthyosaurs, these large shastasaurid suction- or ram-feeders and toothed predators were rather elongate in the body between the pectoral and pelvic fins. Thinking of the proverbial images of sperm whales battling giant squid at ocean depths, it might not be that far off the mark to substitute a shastasaurid for the sperm whale for an equivalent Mesozoic scene. The skulls of *Shonisaurus* and *Shastasaurus*, after all, appear to have been in the range of 2–3 m (6–10 ft) in length. And as we have seen in the previous chapters, the fish and cephalopods were certainly there in the oceans to be fed on, although the presence of large squids, while possible, cannot be confirmed in the Late Triassic; squid-like octopus relatives from the Cretaceous that reached giant squid-range sizes (*Tusoteuthis*) do indicate that such large cephalopods were around during at least part of the Mesozoic, however. So, there may be even more ecological diversity in ichthyosaurs than we realized even just a few decades ago, and their parallels as a group to some marine mammals may be more extensive than just body shape and size in *Ichthyosaurus*.

It is not entirely clear exactly where among the reptile tree the ichthyosaurs originated. At this point, it seems almost equally possible that the first ichthyosaurs evolved from a common ancestor with the aquatic basal reptiles known as mesosaurs, which were mainly around in the Permian, or that they are related to the procolophonids, pareiasaurs, and turtles. We don't yet know.

Meuse Lizards

First reported near Maastricht, Netherlands, in the 1760s and named after a Latinization of the name of the Meuse River (which flows through France, Belgium, and the Netherlands), mosasaurs are unique marine reptiles that ruled the seas of the Cretaceous (figs. 8.2*B*, 8.2*C*, and 8.3). Probably most closely related to varanid lizards, and especially to an extinct group of semiaquatic varanids known as aigialosaurs, the mosasaurs evolved into large predatory marine lizards. They evolved around the same time that ichthyosaurs disappeared. As a group, the mosasaurs were mostly large predators of fish and cephalopods, but there was some variety—*Dallasaurus* was only about 1 m (3 ft) long and had less aquatically specialized limbs, and *Globidens* had blunt, rounded teeth that indicate a diet based on crushing mollusk shells. Among the largest mosasaurs, *Mosasaurus* itself (fig. 8.3A) grew up to 15 m (50 ft) long. These Cretaceous predators were effectively massively overgrown Komodo dragons with flippers physically adapted to prowling the shallow seas.

The largest varanid lizards, Komodo dragons serve as a good model to modify to envision a mosasaur. If a Komodo is 3 m (10 ft) long, our first exercise is to scale it up to, say, *Mosasaurus*'s 12–15 m (40–50 ft). Other mosasaurs were more in the range of 9 m (30 ft), but we'll opt for maximum steroids for now. We then need to shorten the limbs and elongate the fingers and toes into pectoral and pelvic fins. Within our now-established longer length, we also need to play with proportions a little—the trunk needs to be relatively longer, spacing the pectoral and pelvic fins out just a bit, and the neck needs to be a little shorter—and the body and neck may need to be a little meatier. We then need to take this transmogrified Komodo dragon, bend the tip of the tail down slightly, and grow a crescent-shaped, vertically oriented caudal fin. Now just a few tweaks like conical teeth and a snakelike double-hinged jaw joint, and *voila*! *Une* mosasaur.

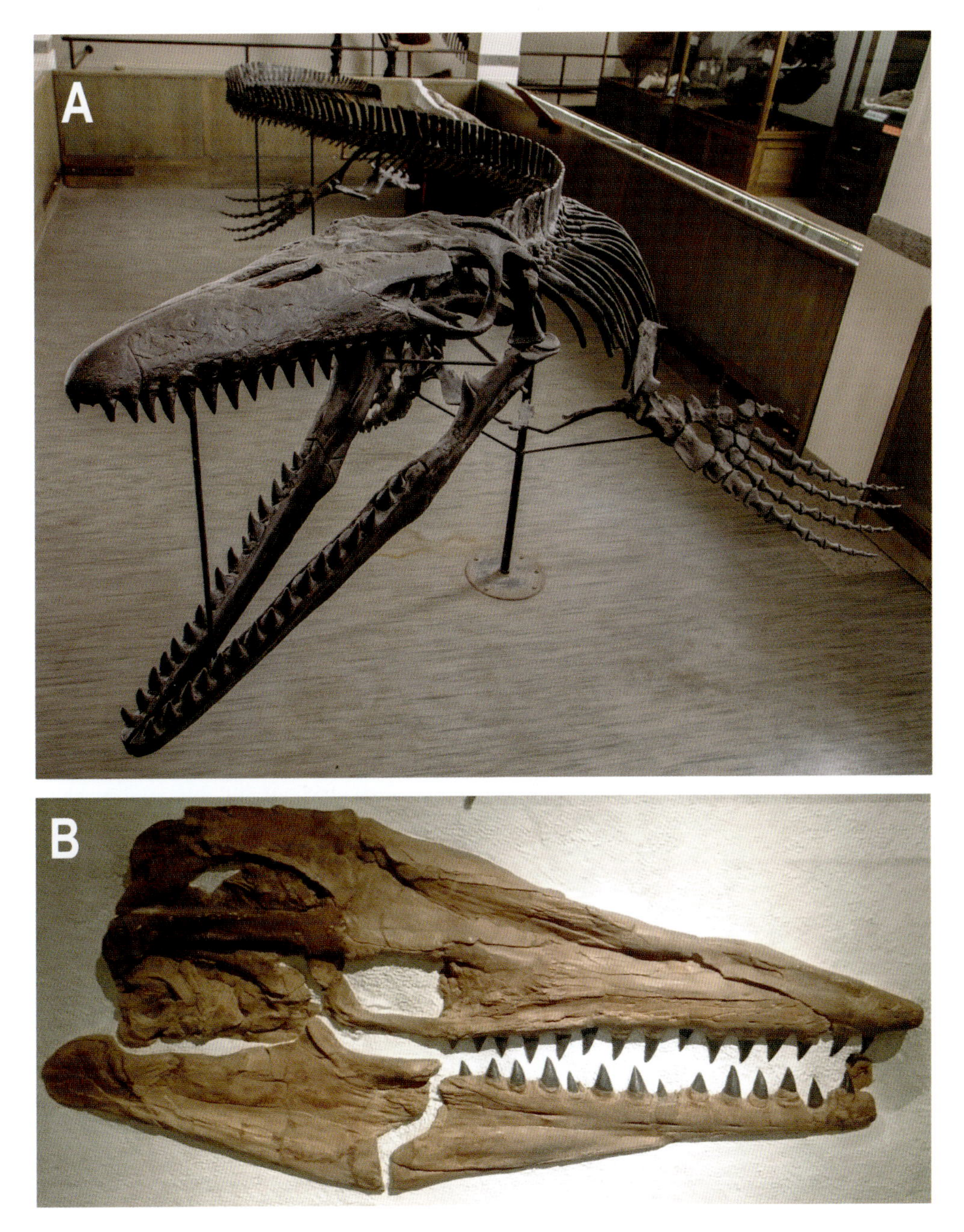

8.3 Some mosasaurs of the Cretaceous. (*A*) Late Cretaceous *Mosasaurus* from the Pierre Shale of South Dakota (SDSM 452). This specimen is about 8.2 m (27 ft) long. (*B*) Cast of the skull of the mosasaur *Tylosaurus* from the Late Cretaceous–age Niobrara Formation of Kansas (UCM 17475). *Photo in A by Kevin Eilbeck, South Dakota School of Mines. Photo in B courtesy of Dean Lomax.*

Mosasaurs were about the closest nature has ever come to producing the giant sea serpents of sailors' nightmares. But they were terrors, regardless, to the seabirds, sharks, fish, sea turtles, cephalopods, and other species of mosasaurs that have been found as stomach contents in some fossil specimens, such as a *Tylosaurus* (figs. 8.2*B* and 8.3*B*) from South Dakota and a *Prognathodon* from Alberta. And we know ammonoids fell victim to mosasaurs also, based on bite marks on fossilized ammonite shells. Other recent finds have indicated that mosasaurs had keeled but otherwise snake-like scales along their dorsal surface and similar but smoother scales along the ventral surface; they were also probably countershaded dark on top and lighter along the belly, as with many modern oceanic species like sharks. It also appears, based on finds of pregnant females from near the aigialosaurs-mosasaurs grade split, that mosasaurs gave birth to live young at sea (fig. 8.2C), as we have seen with ichthyosaurs. And mosasaurs are known from many countries spread across all continents, including Antarctica. This is a testament not only to how widespread the mosasaurs were as a group but also to how high the sea level was in the Late Cretaceous, flooding large swaths of continents and opening up areas with high fossilization potential to the animals.

Unlike the ichthyosaurs, which became extinct in the early Late Cretaceous, right around the time mosasaurs were getting going, the mosasaurs trucked along in all their glory until the massive extinction at the end of the Cretaceous. An unfortunate loss, and one that left sharks as the largest predators in the sea.

Not So Near-Lizards

The first plesiosaurs were found in Europe around the same time that ichthyosaurs of the same ages started showing up in the same rocks. Mary Anning was not done with marine reptiles after working on the ichthyosaur when she was 12. At around age 24, she collected a nearly complete *Plesiosaurus* from the same coastal area near Lyme Regis (fig. 8.4A). This was one of the first specimens to give us an idea of what plesiosaurs really looked like because before this, only isolated bones had been found. Plesiosaurs are marine reptiles with paddle-like limbs, very long to moderately long necks with a small (or sometimes large) skull on the end, and only a short tail. It is likely that they "flew" through the water in a manner similar to sea turtles using both front and back flippers, quite unlike the tail propulsion of the serpentine mosasaurs and the likely very stiff but swift swimming of the ichthyosaurs. They may have had a fleshy fin on the end of the tail to enhance stability rather than provide propulsion. And a few years after *Plesiosaurus*, Mary Anning also found a skeleton of "*Plesiosaurus*" *macrocephalus* (fig. 8.4C), probably a juvenile of an as yet unnamed separate genus. Shortly after this, another "*Plesiosaurus*" showed up, renamed *Thalassiodracon* (fig. 8.4*B*) in 1996. There was no shortage of plesiosaurs in the Jurassic of Europe.

Western South Dakota, the modern version, consists of a sea of auburn grass covering rolling hills all surrounding a conifer-covered island in the plains in the Black Hills and the tips of tan striped mudstone icebergs in the Badlands. The rock on South Dakota's prairie is largely black mudstone, the eroding surface of the Cretaceous-age Pierre Shale. One of the first excavations I was able to help with as a young graduate student was in this

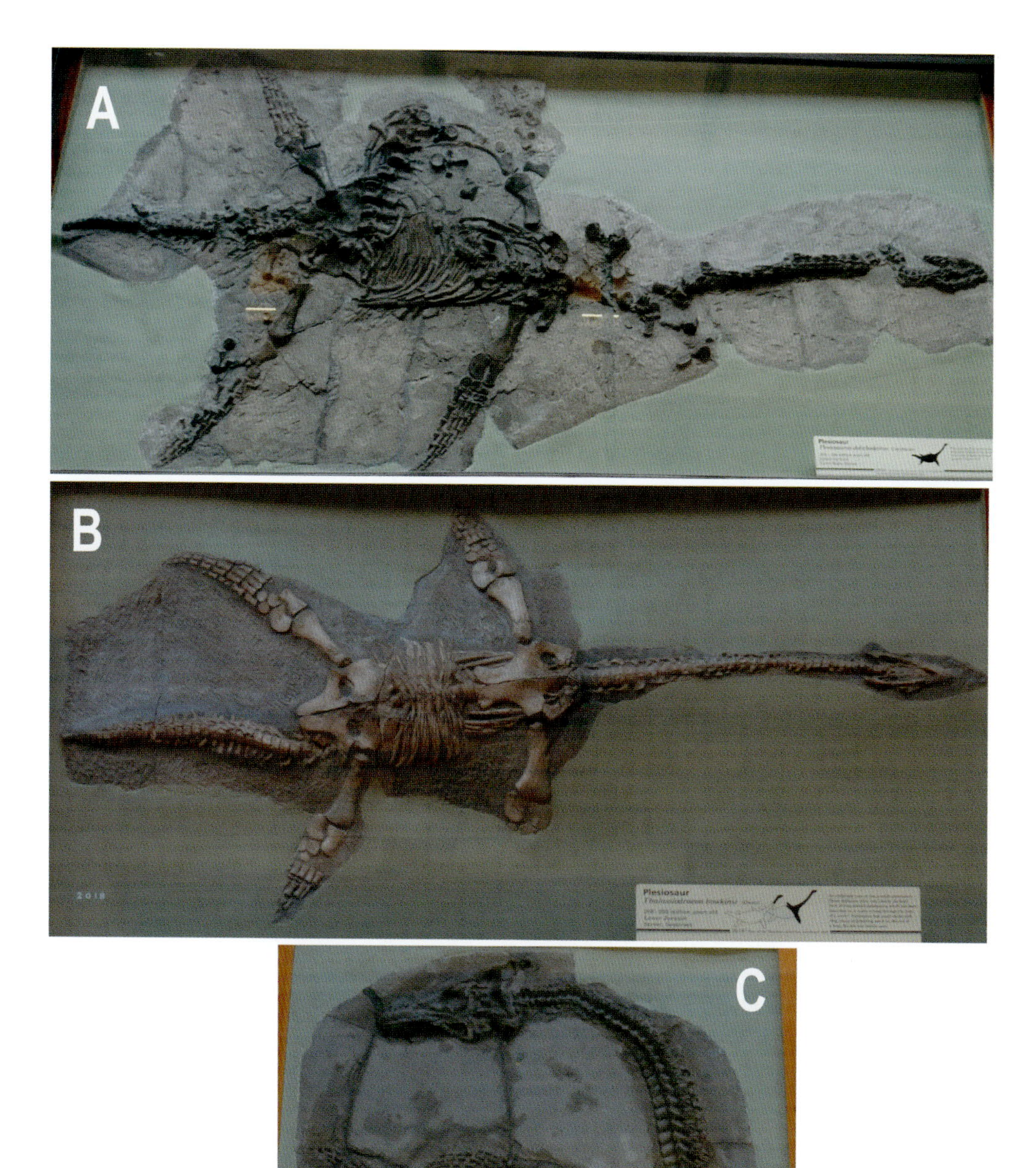

8.4 Early Jurassic plesiosaurs from the United Kingdom, as displayed at the Natural History Museum in London. (*A*) Mary Anning's *Plesiosaurus dolichodeirus* (NHMUK 22656) from Lyme Regis. (*B*) *Thalassiodracon* (NHMUK 2018) from Somerset. (*C*) Mary Anning's "*Plesiosaurus*" *macrocephalus* (NHMUK PV R 1336) from Lyme Regis. *All images courtesy of Dean Lomax.*

terrain, a short drive from Rapid City. The dig consisted of a shallow pit in a low grass-covered hill, but laid out in it were the many lined-up bones of the full flipper of a plesiosaur (in addition to other less dramatic elements of the individual), a resident of the area from about 70 million years ago. The humerus, radius, and ulna, plus finger bones of the flipper were tightly packed and almost perfectly articulated, a well-specialized limb for an aquatic animal. The Late Cretaceous version of South Dakota was part of a seaway stretching from the Gulf of Mexico to the Arctic Ocean, and the black mudstone of today was then the mud at the bottom of that sea, likely several hundred feet down where light was getting to be in short supply. This is where the plesiosaur skeleton had ended up, long before eroding out of that low South Dakota hillside.

Plesiosaurs evolved from more basal pistosaurs such as *Augustasaurus* sometime in the Late Triassic, that time of origin of so many groups that would dominate the Mesozoic and sometimes beyond. Unlike the ichthyosaurs, which didn't quite make it all the way through the Cretaceous, and the mosasaurs, which didn't evolve until around the same time, the plesiosaurs managed to thrive from the Triassic through the end of the Cretaceous, yet another group victim of extraterrestrial bad luck.

Plesiosaurs are divided into two groups: the plesiosauroids and the pliosauroids. The plesiosauroids include, along with a few others, the elasmosaurids, which had very long necks and small heads (figs. 8.2*D* and 8.5), and the polycotylids, which had larger heads and shorter necks. The pliosauroids and most other plesiosaurs other than the elasmosaurids had short (fig. 8.2*E*) to moderately long necks. Some plesiosaur species were as small as 1.5 m (5 ft) in length, but most were a bit bigger, with some species pushing 15 m (50 ft).

The longer neck/small skull versions among plesiosaurs of all groups probably swam more slowly and fed on smaller fish and cephalopods, but the shorter neck/large skull ones were likely active chasers of larger fish and other large prey. Other species appear to have been adapted in their teeth and jaws to strain food out of the water or from bottom sediments. Elasmosaur stomach contents fossils found in Australia indicate that some of these animals bottom-fed on bivalves (partly using gastroliths—stomach stones—to crush the shells) along with some free-swimming cephalopods. Other elasmosaurs ate squid and fish, and possibly more fish as the later part of the Mesozoic went on. Still other elasmosaurs from South America and Antarctica had jaws with "combs" of many long, slender, and interdigitating teeth that did not meet tip to tip; it is hypothesized by plesiosaur researchers such as Robin O'Keefe and others that this tooth arrangement was used to strain food items out of bottom sediment similar to the method used by extant gray whales with their baleen. Conspicuously absent from the Mesozoic, however, would be the role of pelagic, purely plankton-feeding behemoths similar to the baleen whales—at least as far as we know; we may have simply not found those species just yet. Recently found specimens also indicate that plesiosaurs gave birth to live young, and possibly just one per

8.5 The Late Cretaceous elasmosaurid plesiosaur *Styxosaurus* from the Pierre Shale of South Dakota (SDSM 451), as mounted at the South Dakota School of Mines and Technology's Museum of Geology. *Photo by Kevin Eilbeck, South Dakota School of Mines.*

pregnancy, suggesting a high investment in well-developed and large young rather than the opposite strategy of many tiny babies born abandoned and facing a low survival rate.

I once had the pleasure of riding "Pontiac" the mule into the remote reaches of what is known locally as the "North Desert" to collect a large elasmosaurid from the Mancos Shale, northwest of Grand Junction, Colorado. One would think that pack mules would plod along, but apparently, Pontiac was used to leading, because he kept insisting on running to try to pass the other hybrid equine ahead of him. We rode for some time up a remote draw north of "Shale City" (as screenwriter Dalton Trumbo somewhat derisively referred to his hometown of Grand Junction) in the upper part of the Mancos Shale. What we found on the surface of a conical knoll were numerous almost dinosaur-sized vertebrae of the neck and back of a Late Cretaceous elasmosaurid, possibly *Styxosaurus* (figs. 8.2*D* and 8.5). We packed the vertebrae out in the leather saddle bags on the mules, which saved our backs and feet much grief. Some years earlier, and a few miles to the west, near the Colorado–Utah border, a partial skeleton of a smaller shorter-necked plesiosaur had been collected in Prairie Canyon, not far from a bloat-and-float hadrosaurid dinosaur. Plesiosaurs are known also from less remote parts of North America, in the Pierre, Tropic, and Niobrara Formations, for example. But this ubiquity in Cretaceous formations of the region only reflects the fact that plesiosaur material is now known from every continent on the planet and all three periods of the Mesozoic.

Trace borings on plesiosaur bones have also shown us that some modern annelid worms that now specialize in consuming whale bones started out feeding on those of marine reptiles. The strange modern annelids were only discovered in 2002 and were found in the hundreds, bored into the bones of a whale carcass almost 3,048 m (10,000 ft) down in Monterey Bay off the California coast. These worms have no mouths and no guts but still manage to bore into and consume the bone offered by whale falls and that of other vertebrates that also sink to the sea floor. The genus was named *Osedax* (the "bone devourer"), and species have been found all around the world since. It was 13 years later that *Osedax*-type borings were reported from the bones of a plesiosaur (and a sea turtle as well), indicating that these bone-eating worms probably evolved at least by the Cretaceous, feeding on marine reptiles and other large vertebrates during the days of the dinosaurs. And they apparently hung on through the early Cenozoic on whatever large vertebrate falls they could find until dead whales started raining down to the sea floor in the Eocene.

Large Predator Pioneers

These large predator roles seen in the Mesozoic's marine reptiles were largely absent from the ecosystems of Paleozoic seas, except for those played by sharks and placoderm fish such as the nightmarish *Dunkleosteus*. Sharks have been around since deep into the Paleozoic, and they certainly had plenty of representatives through the Mesozoic in both marine and freshwater settings. Placoderms were also significant predators of the seas before the Age of Dinosaurs, but there is no true Paleozoic ecological equivalent of a mosasaur or plesiosaur. And the body shape of the ichthyosaurs was unique at the time it evolved. But more than anything, the Mesozoic marine reptiles demonstrate both the unique (in the mosasaurs) and the familiar, in the convergent body shape and ecology of most ichthyosaurs and dolphins and in the convergent ecologies of the possibly suction-feeding ichthyosaurs and bottom-filtering plesiosaurs (with sperm whales and gray whales, respectively). This will not be the last time we encounter convergence, either. The multiple evolution of similar morphologies and life habits in unrelated taxa has been a theme in Earth's history, and there is yet more to come in our story.

Note

1. Of *Flipper*, Clearwater Marine Aquarium and *Dolphin Tale*, and *Free Willy* fame, respectively.

Age of the Comb Jaws

Choristodera

9

SOUTH OF RANGELY, COLORADO, back in the Upper Cretaceous terrain I worked years ago with Dave Archibald and others, I recently worked a locality our crew had dubbed "Black Widow." This was in a dark gray, carbonaceous mudstone cliff base below a thick channel sandstone in a narrow canyon hidden among the surrounding piñon-juniper benches. This site contained a disappointingly isolated foot bone of an adult duck-billed dinosaur, also known as a hadrosaur, and some turtle material, and was right next to a black widow spider nest in the cliff that we eyed warily for sudden activity as we worked. Luckily, the arachnid stayed put while we removed the bones. But as we excavated there, we turned up an unexpected, small spool-shaped vertebra about the size of a thimble. The dorsal surface of this vertebra demonstrated the distinctive shape of a type of aquatic to semiaquatic reptile known as a champsosaur. Champsosaurs were relatively common in the Cretaceous and made it into the Miocene, but smaller relatives of these later dinosaur-age forms are also known from the Middle and Late Jurassic. The vertebra we found was probably *Champsosaurus* itself. *Champsosaurus* appears to demonstrate differences in the sacrum (or pelvis), suggesting that females were better adapted for moving on land, presumably for egg laying, while males and juveniles may have remained in the water more and were less mobile on land. But the ancestry of these Cretaceous forms, which lived in waterways probably frequented by hadrosaurs and visited by tyrannosaurids and ceratopsians, goes back a ways.

November at 2,134 m (7,000 ft) elevation in Wyoming can be windy and cold. Okay, it's nearly always windy and cold by that point in the fall. It was both a number of years ago when I visited a microvertebrate site in the Morrison Formation north of Como Bluff with Kelli Trujillo of the University of Wyoming. Windy enough, in fact, that Kelli nearly got blown off her feet at one point. A graduate student in paleontology and geology at the time of our trip, Kelli is now a geochronologist at the University of Wyoming and specializes in microvertebrates and radiometric dating of quarries in the Morrison Formation, attempting to unravel the stratigraphic mess of finding out which quarries are older or younger than which. The formation spreads across parts of eight Rocky Mountain states, after all, so one can't just hike along an outcrop to see which site is higher or lower than another on a regional scale. Kelli also specializes in a wide range of acoustic instruments and in beer connoisseurship, but that's when she's not teaching or researching the Morrison Formation. Kelli had found this Morrison site in the course of her dissertation work and was showing me the variety of tiny fossils that had turned up on the surface and through screenwashing.

Among the small vertebrate fossils we found that day were vertebrae of a very small Late Jurassic champsosaur named *Cteniogenys* ("comb jaw"). "Champsosaur" is an informal name based on the Cretaceous–Eocene form *Champsosaurus*, mentioned above, but the formal group name for all of these is Choristodera. *Champsosaurus* was approximately the size of a medium-sized crocodile, with a slender snout and a cordiform and dorsoventrally flattened skull, but *Cteniogenys* from the Morrison was less than 30 cm (1 ft) long in total body length and had a somewhat different skull shape. The skull of *Cteniogenys* was long and triangular with a tapering snout, and the lower jaws were very long and shallow, with tiny conical teeth that had longitudinal ridges near their tips on the tongue-side surfaces. The vertebrae of *Cteniogenys* that Kelli and I found that cold, blustery day north of Como Bluff were about 7–8 mm (0.3 in) in length, which is actually a bit larger than the average size for those fossils in the Morrison—most of the *Cteniogenys* vertebrae I've seen from Wyoming and South Dakota, from numerous localities and in both field and museum collections, are about 5 mm (0.2 in) long. Tooth-bearing skull elements of *Cteniogenys* have been found at a few localities in the Morrison Formation as well, and when working in the Black Hills of northeastern Wyoming, we found a number of jaws in the Little Houston Quarry in what had been an abandoned channel pond; this is where I first became familiar with them. Otherwise, the *Cteniogenys* material in North America is fairly fragmentary. More of what we know of the osteology of the genus is based on elements from the Middle and Late Jurassic of Portugal and the UK. Susan Evans of the University College London has studied much of this material and reconstructed the skull and skeleton, revealing a small reptile with the long skull and possibly a body, tail, and limbs of generally lizard-like proportions. She also found that *Cteniogenys* was likely a basal choristodere and probably an archosauromorph, sharing a common ancestor with crocodiles and dinosaurs but not closely related to either. *Cteniogenys* was a semiaquatic predator that probably hunted small fish and invertebrates in the ponds and possibly the rivers of the Late Jurassic.

Early Diversity

Choristoderes like *Cteniogenys* and *Champsosaurus* are the most common taxa in well-known units in North America such as the Morrison and Hell Creek Formations, but there are numerous other forms from around

***Facing,* 9.1** Choristodera from the Jurassic–Cretaceous of Asia. (*A*) *Coeruleodraco* (IVPP V23318) from the Late Jurassic–age Tiaojishan Formation of China. (*B*) Juvenile of *Monjurosuchus* (IVPP 14261) from the Lower Cretaceous of China. (*C, D*) *Hyphalosaurus* from the Early Cretaceous of the Jehol Biota in China. (*E*) The Early Cretaceous *Ikechosaurus* (IVPP 9611–1). *Images by and courtesy of Wei Gao, Ryoko Matsumoto, and the IVPP (A), and R. Matsumoto and IVPP (B–E).*

A

B

C

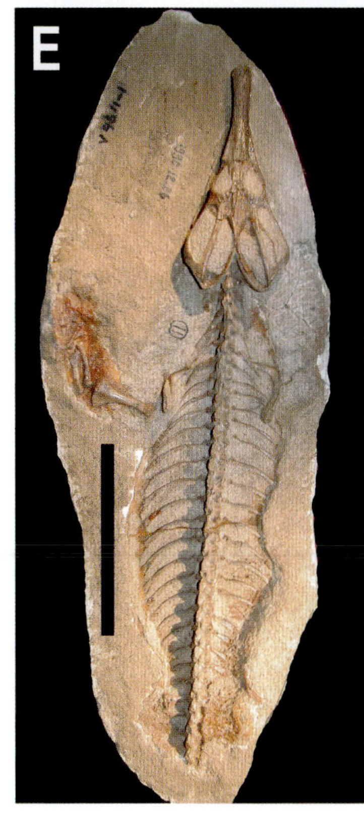
E

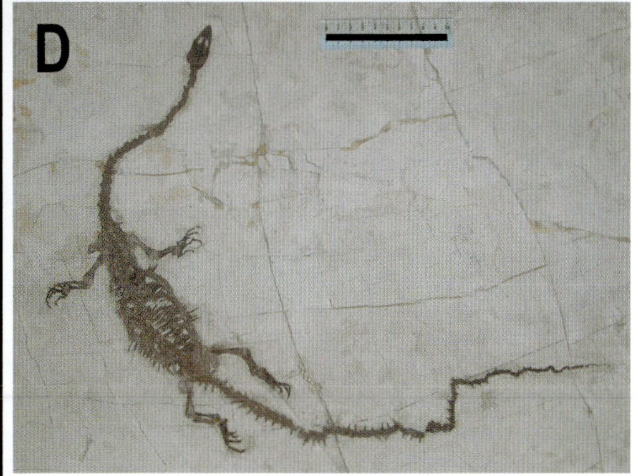
D

the world, including Africa, Asia, and Europe, ranging from at least the Middle Jurassic to the Miocene. Among choristoderes from the age of dinosaurs, *Tchoiria* was a *Champsosaurus*-like form from the Early Cretaceous of Mongolia. *Ikechosaurus* is preserved in several specimens, including a nearly complete skeleton from the Cretaceous of China and Mongolia (fig. 9.1*E*), and *Monjurosuchus* is known from the Early Cretaceous of China and Japan (fig. 9.1*B*). *Hyphalosaurus* from the Early Cretaceous of northeastern China is represented by thousands of specimens, many articulated, with long necks and very long tails (fig. 9.1*C* and 9.1*D*), impressively slender animals sometimes preserved in groups. The known sample of this genus also includes one specimen exhibiting ancient axial bifurcation—a young individual with two necks and two heads splitting from around the shoulders, just like the two-headed snakes you may have seen at the local pet shop (similar to those snakes except for the presence of shoulders and limbs). *Coeruleodraco* was about 40 cm (16 in) long and lived in the Late Jurassic of China (fig. 9.1*A*); it is also far more complete than its near-contemporary *Cteniogenys*. A significant size range is demonstrated by choristoderes, from little *Cteniogenys* up to *Champsosaurus*, with a skull up to about a dozen times longer than its Jurassic cousin.

Choristoderes probably arose in Pangea sometime in the Permian to Triassic. Possible choristodere fossils have recently been preliminarily reported from the Triassic of Germany, in fact. From the Middle Jurassic on, choristoderes seem to be found commonly in wet environments associated with freshwater faunas also including fish, amphibians, turtles, and crocodilians. The last of the choristoderes, that we know of, was little *Lazarassuchus* from the Paleocene to Oligocene of France and Germany and the Miocene of the Czech Republic (fig. 14.4). There are also about four additional genera known from the Cretaceous and Paleocene at various other localities around the world, although they are mostly rare or absent from the southern continents of Gondwana.

Regardless of their size, lizard-like or crocodile-like, choristoderes consistently appear to have been semiaquatic carnivores in the rivers, swamps, marshes, and lakes of their respective environments. Some appear to have specialized in piscivorous diets. Unlike crocodiles, but like mosasaurs and some other reptiles and fish, choristoderes had teeth on the roof of their mouths. In addition to the usual teeth in the jaw bones, palatal bones like the palatine and pterygoid possessed rows or fields of additional teeth in forms ranging from tiny *Cteniogenys* up to *Champsosaurus* and everything in between. Variation in tooth form between species indicates a range of feeding strategies in that some preferred harder or softer prey items. Like crocodiles, they survived the Cretaceous–Paleogene extinction, but it is less clear what did eventually lead to their demise sometime after the early Miocene.

Secret Success

Although they are a small and somewhat specialized group, the choristoderes can be a relatively abundant fossil group in formations like the

Morrison and Hell Creek and in units in China. And their range from the Middle Jurassic to the Miocene, a period of approximately 150 million years, shows that even now-obscure and now-extinct lineages can have been successful for very long periods of time—and should not be accused of being failures or otherwise poorly adapted for their own niches. As I've argued elsewhere before, the real so-called failures don't last long enough to even have much chance to show up in the fossil record in the first place. A group of archosauromorphs that lasted nearly the same length of time as dinosaurs did, and that existed in environments around the world, should not be considered in any way inferior, irrespective of how uncommon they may be as fossils.

10 Something Shocking

Crocodile Rocks

ONE DAY DURING THE SUMMER of 1997, I sat at a lunch counter in a diner in New Haven, Connecticut, with Jack McIntosh, a former Boeing B-29 crew member, retired nuclear physicist, and, as I knew him, sauropod dinosaur guru. As we ate our sandwiches, Jack told me stories of his summer in the 1970s spent driving around the western United States with Kay Behrensmeyer, Robert Bakker, and Peter Dodson and his wife, visiting sites in the Morrison Formation for a project that became "Dodson et al. (1980)," still probably the most-cited single paper on the paleontology of the Morrison Formation after a pair of O. C. Marsh's nineteenth-century installments of the "Principal Characters of American Jurassic Dinosaurs" series. Jack was 50 that mid-1970s summer, and the others were about 30, give or take, so I could only imagine these scenes, and I nearly choked laughing as Jack regaled me with the stories. After lunch, we walked back to Yale University's Peabody Museum, where we were each working on our own projects, for a couple of days. It was not the first time I traveled across the country to a museum somewhere "back East," as we call it, and I happily happened to find myself sharing a collections room with Jack, at work on one of his projects. He still had an office at Wesleyan University, so he spent a lot of time in the big museums with sauropods, so my chances of walking into the collections anywhere along that coast and finding myself blurting out, "Jack!" were pretty good. During that 1997 work at Yale, we went through O. C. Marsh's original Morrison Formation sauropod collections together, and at some point, as we examined the type specimen of either *Barosaurus* or *Diplodocus*,[1] the neural spine simply fell off a caudal vertebra—a moment of mutual horror when Jack and I gasped and looked at each other for a second, thinking, more or less, "Oh crap!" As is standard, we alerted the collections manager to the needed repair before we left but did not attempt to fix the specimen ourselves.

But I was at the Peabody on that trip for more than just sauropods, and I eventually made my way into another part of the basement to spend some time looking at a little skeleton called *Hallopus victor*. This specimen was found in the 1870s near Cañon City, Colorado, and for some time, it was debated whether it was a dinosaur or not and whether it was even from the Morrison Formation or not. As it turns out, it came from high in the Morrison Formation there, and it eventually proved to be a sphenosuchian crocodylomorph—a particular type of small, long-legged, terrestrial stem crocodilian and the first one to appear in the fossil record from the Morrison Formation. Consisting of nearly white fossil bones in a matrix of red siltstone, YPM 1914 (as the *Hallopus* specimen is known) includes a folded-up hind limb of slender femur, tibia, and metatarsal bones plus part of the other hind limb and elements of the pelvis, shoulder, and other skeletal areas. *Hallopus* appears to have had a hind limb significantly longer than the forelimb. There are now, as we will see, at least four types of terrestrial crocs known from the Morrison, but *Hallopus* was the first to be found—another appeared in Wyoming within several years of *Hallopus* but was also thought to be a dinosaur at first, and the others didn't show up until the 1970s and 1980s, just west of the Rocky Mountains.

Interstate 70 starts or ends, depending on which direction you are traveling, near one of the epicenters of so-called power in the country, just outside Washington, DC, and its other end is almost literally in the middle of nowhere, where it T-junctions into Interstate 15 in central Utah at an interchange with no obvious name or any civilization. Traveling west from DC, one passes through Pennsylvania south of Pittsburgh, through the southern parts of Ohio, Indiana, and Illinois, then across Missouri to the Great Plains. Within a couple of days, you pass through the Mile High City and the type section of the Morrison Formation and begin climbing into the Rocky Mountains. Passing 14ers,[2] turnoffs for Breckenridge, and eventually Vail, depending on the season, you are likely to share the road with skiers or mountain bikers on their way to or from downhill banzai runs. If you go in winter, you are likely to end up sitting in traffic in a blizzard with the interstate closed due to too many vehicles having slid off the snow-covered roads. You might even get to witness an avalanche come down and hit the highway. But eventually, you will be on the downslope, and about 227 km (140 mi) west of Denver, you will barely notice a rather unremarkable mountain stream as you cross over it when it appears out of a canyon from the north. You are more likely to miss this stream as a result of watching the Eagle River, which you have been paralleling for almost forty miles now, but you probably don't notice either that the Eagle River is now done and you are now paralleling the stream that just joined you—the Colorado River. The Colorado has just come from its headwaters near Rocky Mountain National Park and Grand Lake, crashing through some narrow canyons that few but train travelers see and emerging onto this well-traveled interstate highway path at a place called Dotsero. From this point, it is just a few miles downstream to Glenwood Canyon, a vertical-walled gorge cut by the river through Paleozoic limestones and Precambrian metamorphic rocks. Here, interstate travelers have to keep one eye on the road while splitting the other's time between the canyon walls and the plunging and, in springtime, raging river. At least riders on the California Zephyr, which also uses the rails going through the canyon, don't have to watch the road. The walls of the canyon are so steep and so close to the river at this point that it took until 1979 to complete the interstate through this section, the last on I-70 to be finished. And during the occasional springtime thaw, house-sized boulders break loose from the cliffs, free-fall, and crater through the road so completely you could drive a van into them and fall to the deck below. But when you emerge from this last real bit of mountains, you find yourself in Glenwood Springs, where Doc Holliday succumbed to tuberculosis in 1887 and where they soon after started on the historic hotel and hot springs pool that now sit right next to the interstate.

In another 80 miles or so, you cross the Colorado River one final time, and then both it and the interstate flow into the Grand Valley, bordered on the north by the gray slopes of the Book Cliffs and on the south by the Uncompahgre Plateau. If you glance south at the right moment as you first enter the valley, you can, on a clear day, catch a glimpse of Mt. Sneffels (4,316 m, 14,158 ft) and the western San Juan Mountains in the vicinity of Ouray, Telluride, and Ridgway. But as the valley opens up, you enter Grand Junction, a small city just under four hours west of Denver in dry, summer nontraffic conditions and only 30 miles from Utah. The town was named after the confluence of the Colorado (at the time, the Grand) and the Gunnison Rivers, which is located just south of downtown. From here, the Colorado flows along the north edge of the Uncompahgre uplift and as a consequence along the south side of the valley before turning into the canyons south and west of Fruita and flowing in isolation into Utah (just south of the Mygatt-Moore Quarry, incidentally). From there, it passes through Westwater Canyon and Fisher Valley and eventually cuts right across Spanish Valley near a sleepy little former uranium mining town called Moab. Sleepy to everyone but the full-time residents of a town of 5,000 that sees two million visitors per year. More quiet miles bring the river below Deadhorse Point, through Canyonlands National Park, and to Cataract Canyon before it slows down and stops for Lake Powell, its brief pause before bombarding through the Grand Canyon.

But Grand Junction is where the Colorado River first enters the open valleys and broad vistas of the Colorado Plateau province. Here, it begins really cutting down into the soft sandstones and shales of Mesozoic such as the Chinle Formation of the Triassic, the Morrison Formation of the Jurassic, and the formations of the Cretaceous, the units that expose the times of dinosaurs in North America. What the Colorado helped expose near Grand Junction was some of the most important outcrops of the Morrison Formation in the region. In 1900, the Morrison here yielded the first-ever *Brachiosaurus* sauropod skeleton. In 1901, Elmer Riggs, the Field Columbian Museum paleontologist who collected that specimen, returned with a small crew and excavated an apatosaur specimen that led him to eventually conclude that *Apatosaurus* and *Brontosaurus* were two names for the same genus of dinosaur—something he recognized as early as a few years after the excavation. And in the 1970s, a new site in the Morrison Formation near Fruita proved to be in some ways even more important.

Fruita Paleontological Area

In March 1977, a meeting was held in Grand Junction. The topic was what to do with a paleontological resource that had been located in the Morrison Formation just two years prior. At the meeting was a disparate conglomeration of researchers, mostly paleontologists. There was George Callison, a 34-year-old reptile biologist from Cal State Long Beach, who had led the crew that found the site. There was Peter Robinson, a mammal paleontologist with international field experience from the University of Colorado Boulder.[3] There was George Gaylord Simpson, a preeminent paleontologist who, at 74, had been specializing in fossil mammals for nearly half a century. There was Russ King, park paleontologist for Dinosaur National Monument at the time, and Nicholas Hotton, dinosaur paleontologist for the Smithsonian's National Museum of Natural History. And there was Lance Eriksen, a Harvard-trained paleontologist who had just recently started working for the institution that is now the Museums of Western Colorado in Grand Junction. They and the Bureau of Land Management were hoping to protect the area and needed to figure out the best course of action.

In 1975, George Callison led a crew out to Fruita to look for lizard and snake fossils in the Morrison Formation. The crew included Jim Clark, now of George Washington University, and Mark Norell, a recently retired senior curator at the American Museum of Natural History. Both were barely in their twenties; Callison was a decade or so older. George was trying to find an area that might preserve small vertebrates of the Morrison and preserve them as more complete skeletons than had been found previously. At the time, just about the only microvertebrates known in the Morrison Formation were from the site known as Reed's Quarry 9 at Como Bluff, Wyoming, but they consisted entirely of isolated teeth, jaws, and bones. George was hoping for more complete material, and, ultimately, things that might offer insights into the origins of snakes, the latter having not been found in the Morrison. As it turned out in the long run, he may have found all three.

But in 1975, the field season for Callison, Clark, and Norell did not start out with any promise of that eventual success. Camping at Colorado National Monument and working a square-mile-plus of Morrison Formation outcrop southwest of and across the Colorado River from Fruita, just west of Grand Junction, George's crew saw little for several weeks. It was the proverbial last-minute find that Jim Clark (fig. 10.1A) made that saved the day and led to decades more work and a significant increase in the knowledge of the Morrison Formation's vertebrate fauna. That find was near the end of the field season, and it consisted of a string of crocodylian vertebrae preserved in a fine sandstone in the lower Brushy Basin Member of the Morrison Formation. This was the beginning of a series of finds of taxa not just new to the Morrison Formation but also better preserved than anything that had been found before in that unit. In addition to the tiny crocodile (which we will get to in a minute), there ended up also being articulated lizards and rhynchocephalians; articulated mammals, including an all-new group of burrowing and termite- and ant-eating mammals (*Fruitafossor*) and several other new species; a large, new species of herbivorous rhynchocephalian, *Eilenodon*; a possible snake; and the world's smallest adult omnivorous dinosaur, *Fruitadens*. Around the same time as these microvertebrates were being revealed, dinosaurs were being discovered in the same rocks just yards away. Specimens of the spiked-and-plated *Stegosaurus*, the giant-blade-toothed carnivore *Ceratosaurus*, the abundant and spoon-toothed *Camarasaurus*, and the common carnivore *Allosaurus* have been found or are still exposed within the confines of the Fruita Paleontological Research Natural Area (FPA), designated such by that 1977 meeting. All this was set off by Jim Clark's find in 1975.

It was too late in their trip for the group to do much with this initial find, but the next year when they returned, they traced the layer to the south and leveled off a ridge above it with a bulldozer (rare that you can do that anymore). This allowed the entire layer to be accessible under only a couple of feet (10s of cm) of mudstone rock at this site, Callison's Quarry. George worked this and another site from the same level just north of Jim's original find (Tom's Place) off and on for the next few years, through the rest of the 1970s and into the '80s, sometimes with Earthwatch groups. For years after that, the site was worked by volunteers Wally Windscheffel, Chuck Safris, Dick Peirce, and others, first for George's permit (with the Natural History Museum of Los Angeles County—LACM), then for the Carnegie Museum of Natural History, and eventually for the Museum of Western Colorado. Tom's Place was later worked by Gabe Bever's and Anjan Bhullar's teams from the American Museum of Natural History and Yale Peabody Museum. I worked at the Callison Quarry off and on for nearly a decade (2004–2013); the site was only ten minutes from my office at the Museum of Western Colorado.

The microvertebrate sites at the Fruita Paleo Area are not the densest deposits of small bones you will ever see, and you can go anywhere from five or ten minutes up to two hours or more without seeing anything but mudstone and pebbles. But you have the chance to see articulated or associated skeletons in the rock, which is what keeps you going just one more split, often for days on end. Usually, your patience is rewarded every two hours or less, if you're lucky, by an isolated find of a lizard vertebra or small limb bone. Sometimes you may even see an isolated mammal maxilla with a few upper molars. But you keep going, waiting for the articulated or associated material.

Through all that digging and waiting, it took me until 2013 to find one of Jim Clark's crocodiles. The species is now known as *Fruitachampsa callisoni*, "Callison's Fruita crocodile." I managed to find just a left dentary (lower jawbone), a tooth, and several associated limb bones, but most of the skeleton is now known from a composite of a number of partial skeletons and numerous skulls from several sites in the FPA. And the LACM now has on display an articulated and nearly complete skeleton of a *Fruitachampsa*.

Wonky Cats: Terrestrial Crocodylomorphs

Fruitachampsa had a short, boxy skull with a short snout and large eyes (fig. 10.1*B*). The external surface of the skull bones and (oddly enough) the palate or roof of the mouth were heavily sculptured with a snaking pattern of pits and ridges (fig. 10.1C). The tooth rows had single upper and pairs of

10.1 Western Colorado and its Morrison Formation shartegosuchid crocs. (*A*) George Callison (*left*) and Jim Clark (*right*), discoverers of the Fruita Paleontological Area (FPA) in 1975, at the FPA's Callison (Main) Quarry in 2010. (*B*) Dorsal view of a skull of *Fruitachampsa* (LACM 128306), a shartegosuchid first found at the FPA. (*C*) Ventral view showing the sculptured palate of another skull of *Fruitachampsa* (LACM 120455), also from the FPA. (*D*) Left lateral view of a small shartegosuchid from the lower Morrison Formation near Delta, Colorado (MWC 5163). Scale bars = 1 cm. *Photos by author.*

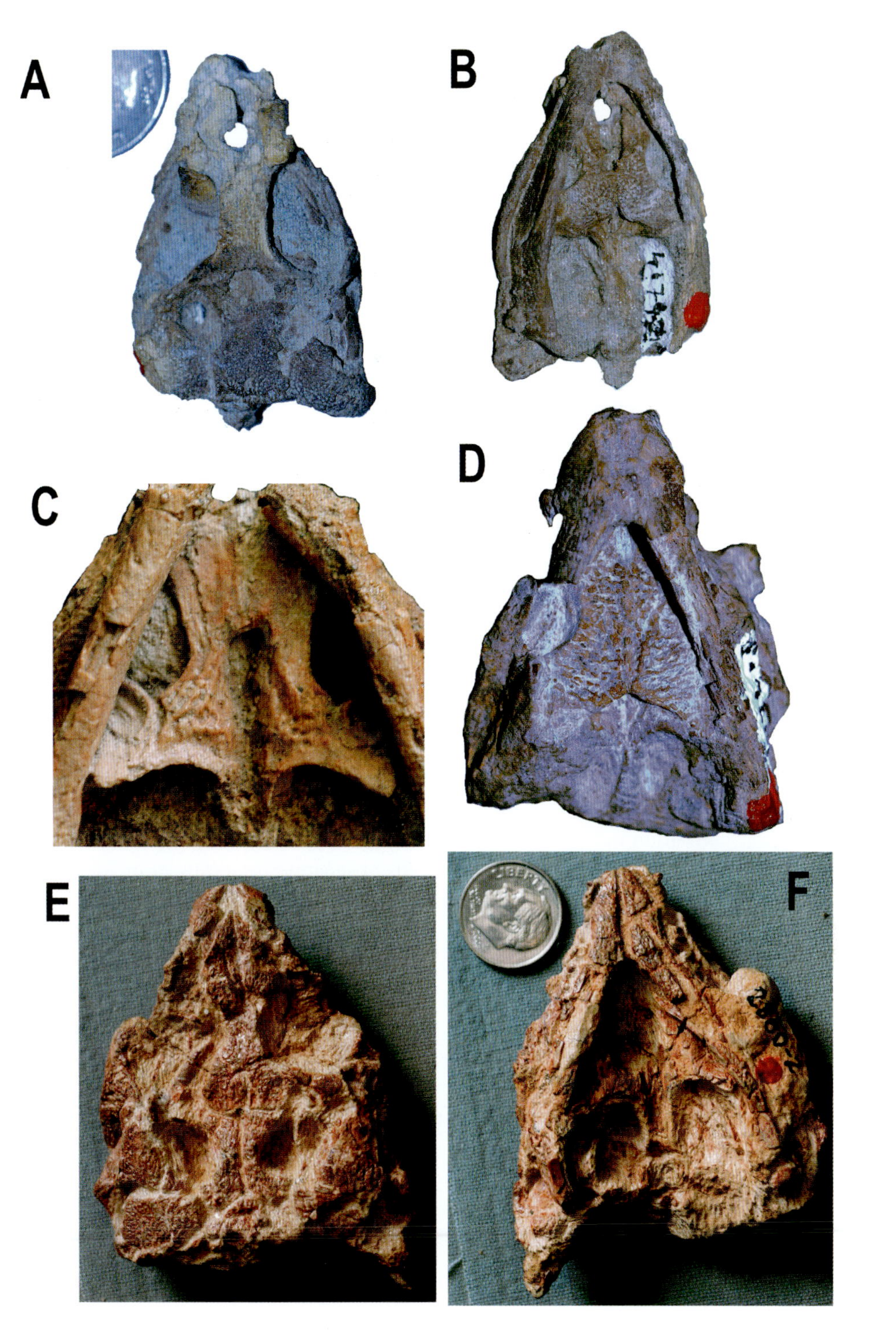

10.2 Some shartegosuchid crocs of Asia. (*A*, *B*) Skull of the Late Jurassic *Shartegosuchus* (PIN RAS 4174/2) in dorsal (*A*) and ventral (*B*) view, Tsagaantsar Formation, Mongolia; length of skull ~4 cm. (*C*) Palate of *Nominosuchus* (PIN RAS 4174/4), Tsagaantsar Formation; full length of skull ~6 cm. (*D*) Ventral view of skull of Late Jurassic *Adzhosuchus* from Mongolia (PIN RAS 4174/5). (*E*, *F*) Skull of the Early Cretaceous *Tagarosuchus* (PIN RAS 2860/2) in dorsal (*E*) and ventral (*F*) views, Ilek Formation, Siberia (possibly a protosuchian). *All images courtesy of Jim Clark and PIN RAS.*

lower canine-like fangs. The limbs were long, and the bones of the wrist were elongate as well. The full length of the animal was up to 1 m (3 ft), but most individuals were a bit smaller. The overall build, proportions, skull, and size were similar to a scaly, reptilian house cat. In fact, I've sometimes referred to these animals as "the house cats of the Jurassic." Like cats, *Fruitachampsa* was built to run and to hunt down and eat small prey items like lizards, mammals, and anything else it could catch. It may well have raided nests for eggs too.

Fruitachampsa and another as yet unnamed and very similar croc from the older Salt Wash Member of the Morrison Formation are what have been called shartegosuchid crocodylomorphs. They are not quite modern crocodiles, obviously with such different size, ecology, and build, and they appear to be related to several other forms that are known mainly from Asia (fig. 10.2), especially at a site called Shar Teg in Mongolia. Among these are taxa such as *Shartegosuchus* (fig. 10.2*A* and 10.2*B*) and *Nominosuchus* (fig. 10.2*C*). The shartegosuchids range from the Late Jurassic to the Early Cretaceous of Asia. There are at least five genera (with a probable sixth yet unnamed) in the family, and most of the Asian forms are from Mongolia and China.

The second, unnamed Morrison form from the Salt Wash Member (fig. 10.1*D*) is one Jim Clark and I have been working on for some time. It was found by Harley Armstrong back in 1987 in matrix collected from an egg site near the Gunnison River southeast of Grand Junction. The skull is in many ways similar to, but in other key characteristics clearly different from, *Fruitachampsa* from stratigraphically higher rocks in the Morrison Formation in the FPA. The interesting part is that the differences between the two suggest that they were parts of separate dispersals from Asia—two different times that shartegosuchids found their way into the area of Morrison Formation deposition all the way from Asia. Also engaging is that most other taxa from the Morrison that are shared between continents are common to North America and Europe or Africa (specifically, for the latter, a deposit in Tanzania); the connection present at the time would have been between northeastern North America and northwestern Europe (and then in turn to Africa) across what was at the time a very narrow, young Atlantic Ocean. Few lineages in the Morrison are connected with Asia, other than our new friends the shartegosuchids. But mammals, lizards, theropod dinosaurs, sauropod dinosaurs . . . all have genera or at least families that occur in North America, Europe, and Africa. During the Late Jurassic, the distance between islands across the short distance of the northern Atlantic Ocean was much less than the paleo-Pacific Ocean, so it seems unlikely that shartegosuchids would have entered the Morrison area from northwestern North America. More likely, they would have come through Europe from the east. Which makes us wonder where the European shartegosuchids are. If a dispersal from Asia comes through Europe, we should expect relatives of *Fruitachampsa* and *Shartegosuchus*, for example, to be preserved in Europe. But this may be an environmental or preservation bias that is favorably presenting us with these animals where we have seen them so far;

even within the Morrison, shartegosuchids are known only from a restricted area of western Colorado and eastern Utah. And Europe at the time was mostly tropical, with islands and shallow seas, rather different from the contemporary environments in western North America and Asia.

At least two localities in the Morrison Formation in western Colorado, one in the Salt Wash and one in the Brushy Basin, preserve shartegosuchid material associated with eggshell, suggesting shartegosuchids may have occasionally raided the nests of dinosaurs. The eggs at the Salt Wash site have not been identified, but the egg material in the Brushy Basin at the FPA, found by Utah state paleontologist Jim Kirkland, is also associated with abundant bones from *Dryosaurus* individuals of various age classes. Perhaps *Fruitachampsa* would help themselves to a *Dryosaurus* egg breakfast now and then.

Several years after *Hallopus* was found in the Morrison Formation in Colorado, another croc-in-disguise showed up at Quarry 9 at Como Bluff, Wyoming, in a pair of anterior lower jaw fragments. It was some reptile with tooth sockets (missing teeth in the fossil) that occurred only in the middle part of the jaw—no tooth sockets at the front where the left and right dentary bones meet at a loose joint known as the symphysis. Instead, the bone spread out in an almost fan-shaped, flat plate. It was what you might call a "spatula-lipped" reptile. And those lower jaw tips were all that had been found at that point. O. C. Marsh described it as an odd type of reptile,[4] but it was eventually suspected of being a crocodile. And when more material of smaller individuals was recognized among fossils collected from the Fruita Paleo Area (FPA) in the Morrison, it became clear that this animal, *Macelognathus*, was in fact a sphenosuchian crocodylomorph related to little *Hallopus* from the Yale Peabody Museum. Like shartegosuchid crocs, these little sphenosuchians were long-legged terrestrial carnivores (fig. 10.3A) that ran around eating small vertebrates and perhaps raiding egg nests.

Sphenosuchians range from the Triassic to the Late Cretaceous of Asia, Europe, North and South America, and South Africa. The group includes genera such as *Litargosuchus* from South Africa (fig. 10.3A), *Pseudhesperosuchus* from Argentina (fig. 10.3B), and *Junggarsuchus* from China (fig. 10.3C).

The notosuchians, another group of crocodylomorphs, range from the Jurassic through the Cretaceous (and likely beyond) and are less slender terrestrial forms. *Simosuchus*, from the Late Cretaceous of Madagascar, for example, has a short boxy skull with leaf-shaped teeth, a relatively short tail, stout limbs and other postcranial bones, and heavy armor, nearly reminiscent of a crocodilian glyptodont or armadillo. It was only about 1 m (3 ft) long and was likely herbivorous. Other smaller notosuchians also appear to have been herbivorous and possible burrow-dwellers, such as the Early Cretaceous *Malawisuchus* from Malawi. *Araripesuchus*, from the Cretaceous of South America and Africa, on the other hand, was also terrestrial and had a relatively short snout, but it lacked the boxy, pug-nosed skull of the former

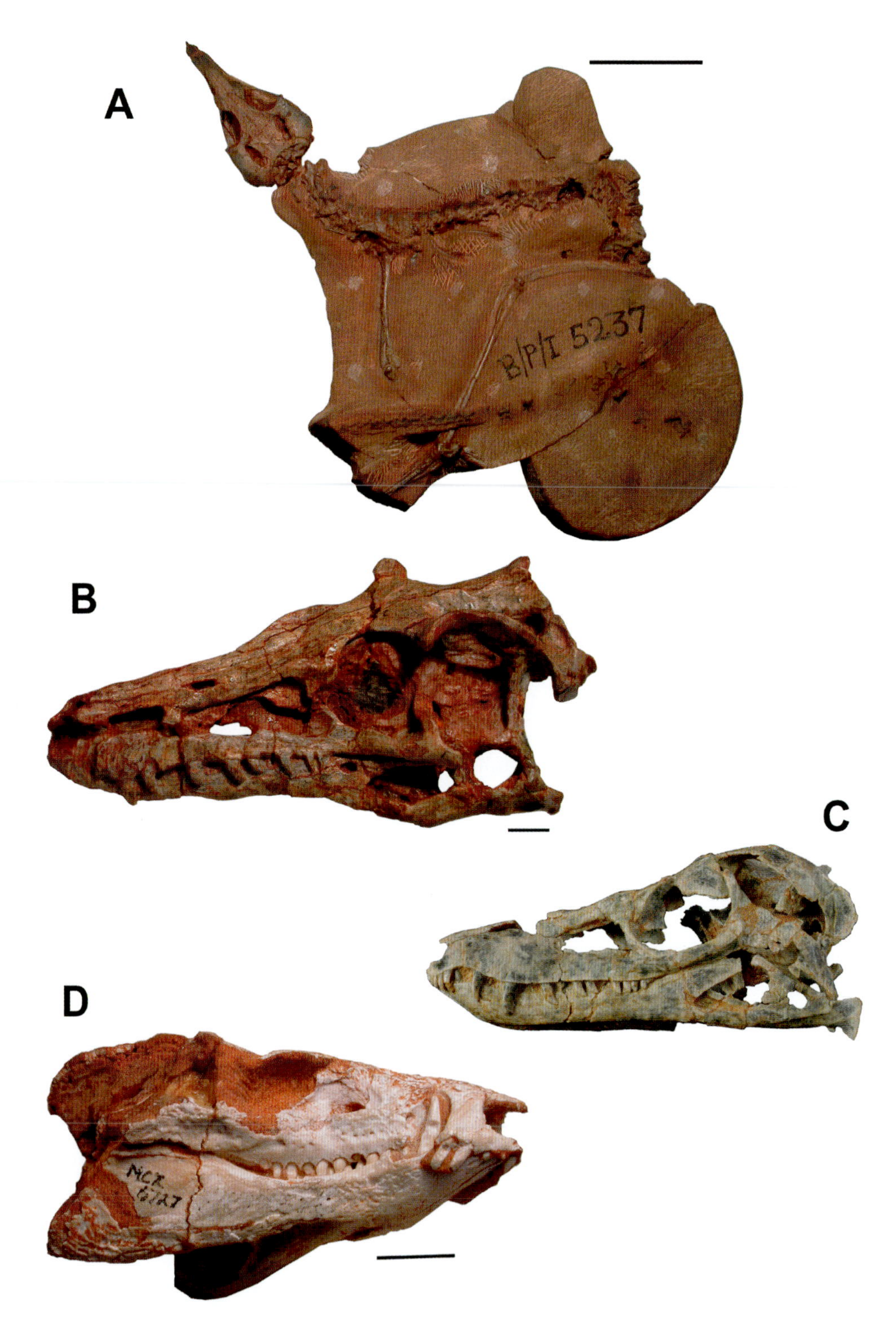

10.3 Some early protosuchian and sphenosuchian crocodylomorphs. (*A*) Skeleton of the Early Jurassic sphenosuchian *Litargosuchus* (BP/I/5237) from South Africa. (*B*) Skull of the Late Triassic sphenosuchian *Pseudhesperosuchus* (PVL 3830) from Argentina. (*C*) Skull of the Middle Jurassic sphenosuchian *Junggarsuchus* (IVPP V14010) from China. (*D*) Skull of the Early Jurassic protosuchian *Protosuchus* (MCZ 6727) from Arizona, US. Scale bars = 1 cm (*B*, *D*) and 5 cm (*A*). *All images from Irmis et al., 2013 (*Geological Society, London, Special Publications *379:275–302) and courtesy of Randy Irmis.*

taxa and had teeth suggesting a carnivorous diet. Paleontologists have so far identified more than 20 genera of notosuchians, some of the least typical crocs one is likely to encounter in the fossil record.

Origins

Small, fast, long-legged carnivores like the shartegosuchids and sphenosuchians of the early millennia of croc history are quite different from the sometimes giant, lazy, sun-lounging crocodilians we know today. Reptilian house cats of the Jurassic bear little resemblance, other than in details of the skeleton, to the massive "ambush logs" that inhabit tropical lakes, rivers, and ponds—or the sand bars around them—in our modern environments. So how did they get from one point to the other?

Some of the answers come from Ward Terrace, the Adeii Eichii Cliffs, and the Painted Desert, down into the Petrified Forest, where the Chinle, Moenave, and Kayenta Formations have yielded much of what we know about North America's earliest crocodiles. Time and again, early representatives of groups are found to originate from this important geographical region. The origins of crocs as terrestrial animals, and their modern condition as mostly freshwater (and in some cases both freshwater and marine capable), oversimplify the picture. Crocodiles have been freshwater almost exclusively since the Miocene, and during the Triassic, they were exclusively terrestrial. During most of the Jurassic and Cretaceous, however, there were terrestrial, freshwater, and fully marine-adapted groups at the same time, and most groups had members that transitioned back and forth, exceptions to every rule, marine and terrestrial specialists appearing from among a group of freshwater forms, for example. Thalattosuchians split from terrestrial crocs by the Early Jurassic and took to the shallow seas (with a few species returning to fresh water later). Freshwater specialists appeared by the Early to Middle Jurassic (then a few became marine). Terrestrial specialists hung in there until the Miocene in the notosuchians (and a few became freshwater). It was an ecological game of musical chairs for the crocodiles of the Jurassic and Cretaceous.

Crocodiles in general (more precisely Crocodylomorpha) appear in the Triassic period, about 230 million years ago, around the same time as dinosaurs and so many others. An example of one of these early crocs is little *Hesperosuchus* from the Upper Triassic Chinle Formation of the American Southwest (Arizona and New Mexico), a taxon about the size of a small-to-medium dog with serrated teeth that indicate a carnivorous diet. Then there was *Sphenosuchus* from the Early Jurassic–age Elliot Formation in South Africa, a form that was up to nearly 1.5 m (5 ft) in total length and was also terrestrial and carnivorous. And it appears to have been related to other later forms, including *Kayentasuchus* from the Lower Jurassic Kayenta Formation in northern Arizona and some other sphenosuchians from the Morrison Formation in Colorado and Wyoming. In fact, most of these early crocodiles would not be particularly familiar to us, as they too were small, long-legged, terrestrial runners quite different from the semiaquatic forms we know today.

These crocodylomorphs were just outside the Crocodyliformes, the latter a more exclusive clade within Crocodylomorpha that contains modern crocs and other ancestors. Among these groups of crocodyliforms are the protosuchids, the atoposaurids, the shartegosuchids that we already met, and the Neosuchia, a yet more exclusive clade containing most large, ancient semiaquatic crocs and the modern Crocodylia. Crocodyliformes is defined as the most recent common ancestor of *Protosuchus* and *Crocodylus niloticus* (the Nile crocodile) plus all its descendants. *Protosuchus* was found in Early Jurassic–age deposits in North America (fig. 10.3*D*) and South Africa, and it had a short, triangular skull (in top view) with a narrow snout, a relatively long trunk and tail, but long legs as well. It also had bones within the skin forming an armor along the back, as is common in a number of crocs. The first *Protosuchus* was found in the Lower Jurassic Moenave Formation of northern Arizona, a western and partly fluvial equivalent to the dune deposits of the Wingate Sandstone of eastern Utah fame. Another possible protosuchian is known from the Upper Jurassic Morrison Formation in Utah in little *Hoplosuchus* (fig. 10.4A).

Neosuchia comprises the crocodyliforms of more "modern" aspect, which means a number of fairly diverse groups, including some extinct lineages, plus the crown-group Crocodylia. Among the primitive neosuchians are the goniopholidids (fig. 10.4*B*, 10.4*C*, and 10.4*E*), a group whose earliest representative may date back to the Early Jurassic in *Calsoyasuchus* from, again, the Kayenta Formation in Arizona. Among the other goniopholidids were numerous Late Jurassic to Early Cretaceous forms from North America (such as *Amphicotylus* [fig. 10.4C], and *Eutretauranosuchus* [fig. 10.4*B* and10.4*E*]) and elsewhere, plus Europe and Asia's *Goniopholis*.

Although goniopholidids were not modern-grade crocodilians, they were more or less ecologically equivalent to modern crocodiles, with similar overall body form and habits. There were at least a few structural (osteological) differences. The skulls were very similar in overall structure and sculpturing to most modern forms but lacked specific details of the prefrontal and pterygoid bones, the latter related to the choanae, the opening between the nasal cavity and the palate on the roof of the mouth. The vertebrae of goniopholidids were flat to slightly concave on each face of the centrum rather than procoelous (socket in front, ball in back) as in modern crocs. Goniopholidids lurked in the often-muddy waters of the lakes and ponds of many dinosaur ecosystems in at least North America, Europe, and Asia from the Jurassic through the Early (and into the Late) Cretaceous. Small dinosaurs were likely in considerable danger from the crocs along the shores of many of these bodies of water for millions of years.

Pholidosaurids ranged from the Middle Jurassic through the rest of the Mesozoic and could get quite large. During the Cretaceous in northern Africa, the pholidosaurid *Sarcosuchus* was a giant croc that may well have fed on dinosaurs, having reached lengths of up to 9.5 m (31 ft). In North America, there is another early Late Cretaceous pholidosaurid named *Terminonaris* that got up to about 5.8 m (19 ft) long and lived along the Western

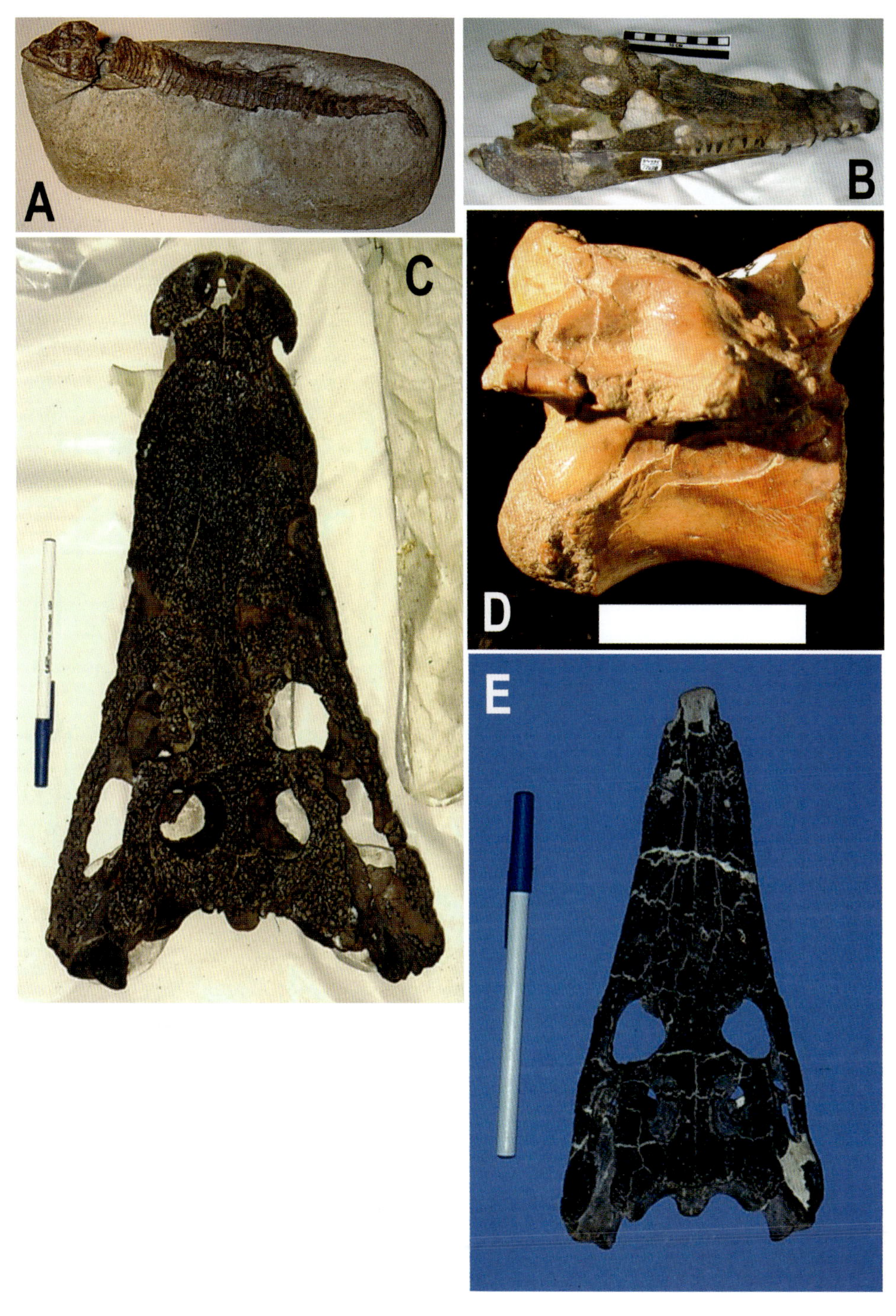

10.4 Some crocodylomorphs of North America. (*A*) The possible protosuchid *Hoplosuchus* from the Upper Jurassic Morrison Formation of Dinosaur National Monument, Utah (CM 11361). (*B*) The goniopholidid neosuchian *Eutretauranosuchus* from the Dry Mesa Quarry in the Morrison Formation. BYU 17628. (*C*) The goniopholidid neosuchian *Amphicotylus* (AMNH 5782) from Colorado's Morrison Formation. (*D*) A small crocodilian vertebra in right lateral view, Upper Cretaceous Williams Fork Formation of Colorado (specimen located at Museums of Western Colorado, MWC 8222). (*E*) The type specimen of *Eutretaura-nosuchus* from the Morrison Formation (Upper Jurassic) of Colorado (CMNH 8028). Total length in *A* ~15 cm, scale in *B* is 10 cm, pen in *C* and *E* ~15 cm, scale in *D* = 1 cm. All photos by author.

Interior Seaway from Texas up to Saskatchewan, with one specimen known from Germany.

The rise of semiaquatic crocodilians in the Early and Middle Jurassic may have resulted in part from the loss of the phytosaurs at the end of the Triassic. Before then, phytosaurs had occupied the role of large, cone-toothed, narrow-snouted semiaquatic predators of all things aquatic and terrestrial-but-thirsty. Once they were gone, a few species of small, terrestrial crocs took advantage of a now-unoccupied niche and began copying the phytosaurs, becoming larger, heavier, and acquiring longer, narrower skulls. Their legs also became shorter relative to the body, and the tails became laterally compressed to help propel them through the water that they now lived in much of the time. Once they appeared and took over the tropical rivers and ponds of the Earth, these crocs lived alongside their small, gangly terrestrial cousins throughout the rest of the Mesozoic and have dominated the semiaquatic role ever since.

Then there were the teleosaurid thalattosuchians (of all tongue-twisting names), a group of marine crocodilians that appear to have lived close to shore, scavenging carcasses of terrestrial animals as well as hunting turtles. There were a number of these forms found in the Late Jurassic of Europe. A sister group to the thalattosuchians was the Cretaceous- to Eocene-age dyrosaurids, which appear to have lived in marginal marine habitats like estuaries and nearshore settings.

Another group of extinct neosuchians is the Atoposauridae, which consists of relatively small, apparently at least in part semiaquatic, crocodilians with relatively long limbs that ranged from the Late Jurassic to the Late Cretaceous, although most are Late Jurassic to Early Cretaceous. This group includes forms such as *Alligatorium* from the Late Jurassic of France and *Knoetschkesuchus* from the Late Jurassic of Portugal and possibly *Theriosuchus* from the Late Jurassic to Early Cretaceous of England, Thailand, and North America. One characteristic of these atoposaurids is teeth that are differentiated in form along the tooth row so that some are round in cross section and small, or round and large, or laterally compressed and medium in size. Paralligatorids include forms from the Jurassic and Cretaceous in Mongolia, China, Brazil, Europe, and the United States. Recent analyses have suggested that atoposaurids and paralligatorids may be basal eusuchians, that is, more derived neosuchians consisting of crown crocodilians and a few extinct forms.

This early diversity brings us back to one of the more stimulating aspects of ancient crocodiles, one that we encountered earlier. They weren't all vicious carnivores. Most were and are, yes, but a number of forms developed more complex teeth and appear to have ventured into omnivory and herbivory. Herbivorous crocs? Yes, stranger things have happened in biology. In Upper Cretaceous rocks in Madagascar, paleontologists uncovered the aforementioned *Simosuchus*, the short-tailed and squat croc with leaf-shaped teeth and abundant armor plating along the back, tail, and limbs, which likely was herbivorous. This was one of the first indications that croc diets were not always what we assumed. There was *Chimaerasuchus* from the Early Cretaceous of China, a crocodyliform with blunt but simple multicusped teeth that caused it to initially be mistaken for a multituberculate mammal! Another apparent herbivore was the "cat crocodile" *Pakasuchus* from the Early Cretaceous of Tanzania, an animal with teeth differentiated, mammal-like, into caniniform, premolariform, and molariform types along the tooth row. The notosuchian *Armadillosuchus* was an omnivore from the Late Cretaceous of Brazil whose armor was so extensive it paralleled that of modern armadillos—or, rather, modern armadillos have copied *Armadillosuchus*.

So crocs have not always been large, semiaquatic, lazy-most-of-the-time carnivores. Many in the past have been small, terrestrial runners that could be, in some species, omnivorous or herbivorous. Yet, by the Middle Jurassic, the large, semiaquatic lineages had appeared and become much more diverse and abundant, but they did not replace the terrestrial types that represented Crocs 1.0. Instead, for some time, the two ecologies coexisted in the same times and regions, if not necessarily exactly the same subenvironments. By the time of the Morrison Formation in the western part of North America, about 150 million years ago, there were (that we know of) at least eight different types of crocodylomorphs living in what is now the Rocky Mountain region. Five of these eight were small, long-limbed, catlike predators of the land: the shartegosuchid *Fruitachampsa*; the sphenosuchians *Macelognathus* and *Hallopus*; the possible protosuchid *Hoplosuchus*; and the as-yet-unnamed shartegosuchid that Jim Clark and I have been working on for nearly a generation. Another two were large, short-limbed, semiaquatic ambushers of fish and anything that got too close to the water, the goniopholidids *Amphicotylus* and *Eutretauranosuchus*. *Theriosuchus* was a small atoposaurid and may have been semiaquatic or more terrestrial, we're not sure, though its fossils are associated with wet paleoenvironments. If you lived in the Late Jurassic in North America, you would be in danger in and near the water, and your smaller pets would be in danger from the terrestrial shartegosuchids and sphenosuchids.

A broadly similar cast of characters was around Europe in the Early Cretaceous, except for the shartegosuchids and some of the other small crocs. In the Wealden Group of England, there were about 11 species of crocs in nine genera, including goniopholidids and atoposaurids.

Modern Forms

Finally, by the early part of the Late Cretaceous, Crocodylia itself appeared, the order of crocodylids, alligators and caimans, and gharials. Outside Rangely, Colorado, during our searches of the Williams Fork Formation that summer with Dave Archibald, we saw plenty of crocodilians in with the turtles and fish and other fauna of that coastal delta deposit. We continue to see them on almost every visit to the field area. Teeth and osteoderms, the armor bones embedded in the skin, are most common, but parts of skulls and vertebrae (fig. 10.4*D*) show up occasionally too. Most of these probably belong to crocodiles such as *Leidyosuchus*, but also present are alligatoroids like *Brachychampsa*. It is hard not to wonder if such settings

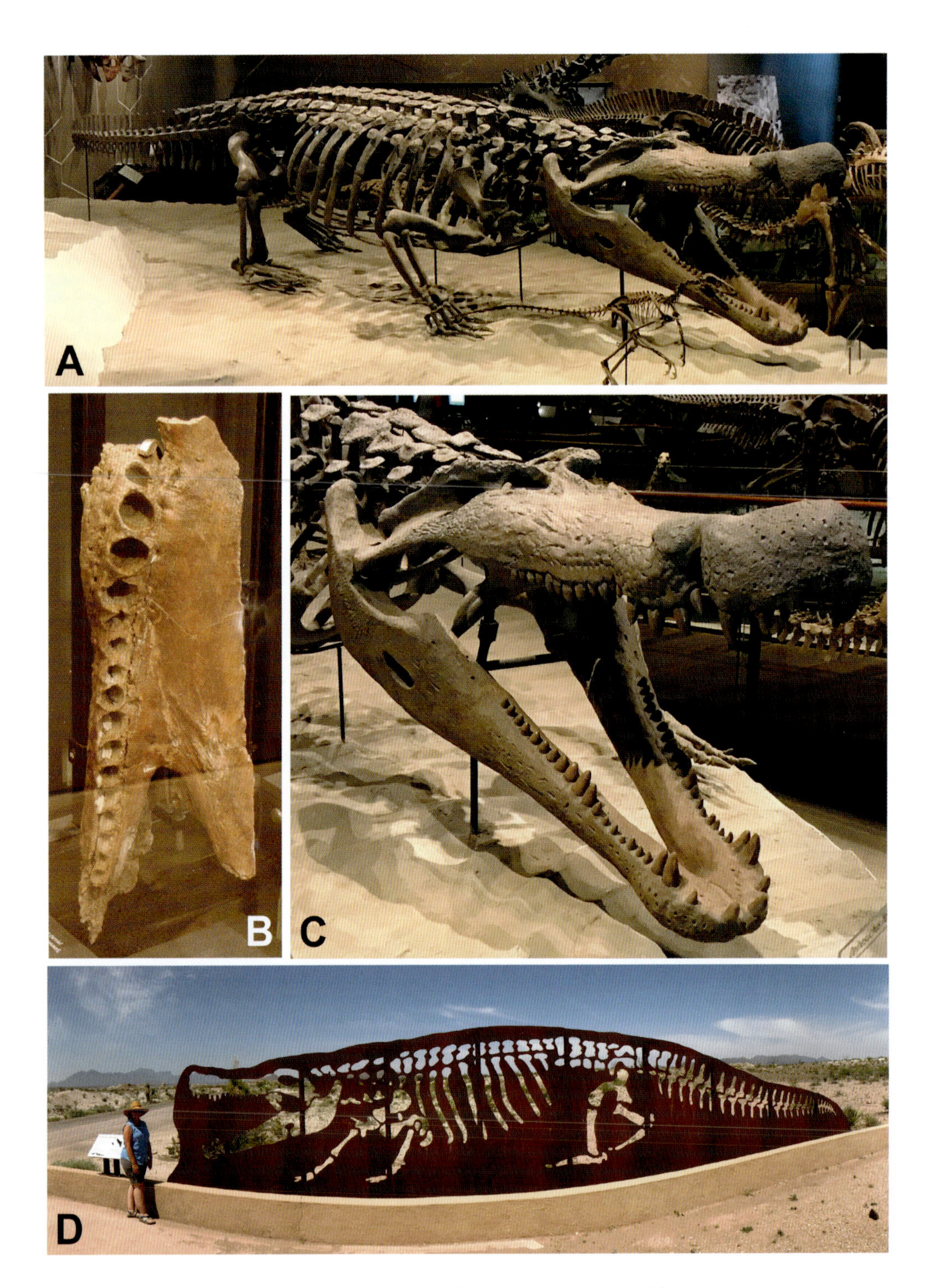

10.5 The giant Late Cretaceous alligatoroid *Deinosuchus*. (*A*) Skeletal reconstruction at the Natural History Museum of Utah. (*B*) Right maxilla of *Deinosuchus* specimen from the Kaiparowits Formation of southern Utah (UMNH VP16783). (*C*) Detail of the skull of *Deinosuchus* in same mount as *A*. (*D*) Reconstruction at Big Bend National Park in Texas, showing size of largest specimens of *Deinosuchus*; 1.7-m (5.6 ft) paleontologist for scale. *All photos by author and A–C courtesy of the Natural History Museum of Utah.*

along the shoreline, and the animals that lived in them and that we find on almost every visit, ever experienced Western Interior Seaway hurricanes such as the Mississippi Delta does today. The crocodile and alligator taxa known from the Williams Fork are also found in other areas of the modern Rocky Mountain region, both north and south, throughout the rest of the Late Cretaceous. For the next five million years, descendants of these crocs occupied rivers and swamps along the shoreline of the interior sea, and *Brachychampsa*, for example, is known from the Hell Creek Formation at the end of the Cretaceous, along with the likes of *Tyrannosaurus* and *Triceratops*.

Deinosuchus was an alligatoroid that lived in coastal freshwater and marine environments along the east coast of Laramidia and the southwestern to southeastern shores of Appalachia during the Late Cretaceous (fig. 10.5A), essentially at the same time that *Leidyosuchus* and *Brachychampsa* were lurking around the swamps that would become the rock south of Rangely, Colorado. *Deinosuchus* was a reasonably large crocodilian at 12 m (39 ft) total length—okay, it was another giant (fig. 10.5*D*). The skull alone was nearly the size of that of *Tyrannosaurus* at around 1.7 m (5–6 ft) (fig. 10.5C), and it was especially robust, with truly massive, conical teeth. It has been found in Big Bend National Park in Texas, plus Wyoming, Utah (fig. 10.5*B*), and Montana on the western coast of the Western Interior Seaway. Along the southern coast of Appalachia, *Deinosuchus* has been found in Mississippi, Alabama, Georgia, North Carolina, and New Jersey. It appears that the western *Deinosuchus* populations may have consisted of individuals that were, on average, a bit larger than those from eastern areas.

And more than just small dinosaurs were endangered by *Deinosuchus*—probably most dinosaurs of the time could have been ambushed by this predator, with a skull that large and robust. In fact, bite marks of a size and shape matching *Deinosuchus* have been found on marine turtles and hadrosaur dinosaur vertebrae preserved in the same deposits as the croc, and it appears that eastern *Deinosuchus* were more often feeding on turtles, and the western ones more often on dinosaurs, at least based on preliminary evidence. Trips to coastal waters could be dangerous even for hadrosaurs.

Fascinatingly, the last of the ridiculously large crocs didn't actually appear until late in the Age of Mammals in South America, well after nonavian dinosaurs were gone, but many of these aforementioned giants of the Cretaceous were terrifying even among giants.

The Crocodile World

One thing that the crocodiles of the Mesozoic demonstrate, at least the semiaquatic ones that seem ecologically similar to those today, is where the climate was warmer in the past. Crocodilians today tolerate only rare freezes and short ones at that. It generally needs to be tropical for them to thrive. In North America, this means we get alligators and American crocodiles in Florida, Georgia, and Louisiana but not very far north or inland at all. One also sees crocodiles in Central and South America, plus in the Caribbean such as rivers of Cuba and Jamaica. Elsewhere, crocs occur in central Africa and the Nile and the Serengeti, plus in northern Australia and the tropics of Asia. In the past, of course, at various times, crocodilians lived in now-impossible places like Wyoming, Montana, and the Arctic. This is because those first two areas were lower in elevation back then and were often near the inland sea, which helped moderate their climate. Also, in all three examples, the simple truth was that the global climate was warmer with the particular effect that tropical temperatures and precipitation extended much farther north, in the Northern Hemisphere, than they do today. Thus, Wyoming had crocodilians almost continuously from about 160 million years ago until the Oligocene (about 34 million years ago), for most of the Age of Dinosaurs and then some. During the Eocene in Wyoming, about forty-five million years ago, the crocodiles lived in lakes and rivers surrounded by palms, magnolias, ferns, and flowers in a fully tropical climate. Even in the summer in Wyoming, as gorgeous as it is, it is about as far from crocodile country as you can imagine—in the almost constantly lethal winter in Wyoming, you can't imagine much of anything other than getting back indoors. To think that the state was crocodile country for more than 120 million years gives us some perspective on the constant change in Earth's climatic and biologic history.

The first crocodile fossil I ever saw was probably somewhere in the Eocene rocks of Wyoming, out with Don Prothero and Al Tabrum (of the Carnegie Museum of Natural History) or Bob Emry (of the Smithsonian) when I was in college. Since then, I've found or worked on more croc fossils than I can recall, and it is not as if they are my specialty. Far from it. In working on other projects out in the field, I've stumbled on or excavated croc specimens in the Jurassic of the Black Hills; the Jurassic around Medicine Bow, Wyoming; the Jurassic of Utah and Colorado; the Paleocene of North Park, Colorado; the Eocene of Bridger, Wyoming; the Paleocene of Hanna, Wyoming; the Late Cretaceous of northwestern Colorado . . . they're everywhere, even when you're not necessarily looking for them. How much could one find if one were specifically hunting the crocodiles out? A lot more than I have! Although they may be a bit of a "background" fauna, at least in most people's minds, the crocodiles of the Mesozoic are relatively abundant and an important element of the ecosystems they inhabited, and their descendants still are today.

Back in the western Colorado and eastern Utah region, around Grand Junction on that lonely stretch of Interstate 70 we started out on, and almost to its unceremonious termination at Interstate 15 in the middle of Utah, the rocks contain evidence of nearly the entire Mesozoic history of crocodiles and forms of both terrestrial and semiaquatic habit. Just in the Grand Junction area, one can find footprints of the first terrestrial crocodylomorphs of the Late Triassic in the Chinle Formation (near the little village of Gateway in the gorgeous setting of the Dolores River canyon and along the Colorado River in Ruby Canyon); more footprints of similar terrestrial crocs from the Early Jurassic oasis mudflat deposits of the mostly sand dune–deposited Wingate Sandstone (also near Gateway and elsewhere); by the Middle Jurassic, there is little *Entradasuchus*, a crocodyliform that is the only vertebrate body fossil from the entire Entrada Sandstone (this site is north and west of Gateway); by the Late Jurassic, there are most of the Morrison Formation groups named above, from shartegosuchids and sphenosuchians to goniopholidids, at sites ranging from the Fruita Paleo Area down to Dry Mesa Quarry and out to Rabbit Valley; Early Cretaceous crocs from the Cedar Mountain Formation along the I-70 corridor just over the border into Utah include atoposaurids, pholidosaurids, goniopholidids, a possible teleosaurid, and a small semiaquatic neosuchian called *Bernissartia*; and finally, in the Mesaverde Group and Williams Fork Formation, you get teeth and bones of alligatoroids, crocodiles, and a possible pholidosaurid, plus tracks of giant crocs.

It was a crocodile world back then, you might say. The dinosaurs just lived in it. Is that an exaggeration? Probably. But certainly, the rivers, lakes, and ponds of the Cretaceous were ruled by the semiaquatic crocs, not the fully terrestrial dinosaurs, at the time. And unlike today, the earlier part of the Mesozoic had the terrestrial crocs running around as well, along with some herbivorous and omnivorous forms, some of which extended well into the Cretaceous.

And the crocodile-friendly climatic conditions continued well after the Cretaceous during the age of mammals. The shrinking of their habitats ever closer to the equator really didn't begin until after the Eocene and even then, for a while, you still had some species of crocs in places like South Dakota during the Oligocene. And more of the largest crocodilians ever to have existed were found in South America in Miocene deposits.

Lucky for us, the crocodilians continue to thrive in modern environments, although we are perhaps also fortunate that things like *Deinosuchus* and *Sarcosuchus* are not around anymore. The fossil evidence of those forms will suffice.

Notes

1. I can't remember which all these years later—but something collected in either 1889 or 1877.
2. Peaks higher than 14,000 ft (4,268 m).
3. Peter was still there 20 years later when I was a grad student there.
4. It was odd, and possibly a new order, according to Marsh.

Wing Fingers

Pterosaurs

11

IN THE CENTER OF WHAT REMAINS of old town Munich, Germany, on the side of the New Town Hall facing Marienplatz (center of Munich since the twelfth century), is the Rathaus-Glockenspiel clock tower, which puts on a show of two folktales, two to three times a day, with life-size human figures on revolving platforms in a type of elaborate cuckoo-clock way of marking certain hours. This spectacle attracts tourists and occasionally locals to this heart of the city. Not far away from Marienplatz are the double towers of the Frauenkirche, a church and icon of the city dating from the fifteenth century. Several blocks away in a different direction is another tourist draw in the Hofbräuhaus, which originated in the sixteenth century as a brewery and only became the attraction it is today after World War II. Years ago, my cousin and I were among the tourists visiting these sites, multiple times on several visits to the city, during a summer spent train-hopping around the Continent, and we also spent a couple of days at the Deutsches Museum, situated on an island in the middle of the Isar River near the center of the city. This is the largest museum of science and technology in the world, and we didn't come close to seeing all of it, but it was impressive, and I, of course, was drawn to the handful of paleontology exhibits there—Paleozoic tree stumps, ichthyosaurs, and a few others. But what I didn't yet realize was that if my cousin and I had gone only about one and a half hours by car out of Munich, we would have been in the heart of one of the most important paleontological landscapes in the world.

Solnhofen Lagerstätte

If one drives those one and a half hours northwest of Munich, one comes to Solnhofen, a village on a bend in the Altmühl River, a tributary of the Danube, around which are limestone quarries that produce some of the finest slabs of limestone around. These thin limestone slabs have been mined for centuries for art, roof tiling, lithography, and other uses. But in these limestone slabs occasionally are fossils of such quality that the site is among the best in the world for representing any ecosystem, and it is certainly the best for the Late Jurassic. Downstream is Eichstätt, another town on the banks of the Altmühl, and it has numerous quarries around it as well. The fossils from the Solnhofen Limestone are Late Jurassic in age and are essentially the same age as those from the upper layers of the Morrison Formation in North America, a unit we have visited several times already in this book.

Whereas the Morrison Formation represents mostly flat inland floodplains, the Solnhofen Limestone was deposited in an entirely different setting. During the Late Jurassic, Europe was a mixed setting of islands and shallow tropical seas. The open Tethys Sea was to the southeast, and much of the continent was flooded by a shallow sea. The area of the Altmühl valley north of Munich had several relatively deep, restricted basins of lime mud surrounded by shallow shoals of coral reefs and algae- and sponge-based mounds. Islands existed to the northeast and northwest. The deep basins between the mounds and reefs of the shallow seas may well have been anoxic and highly saline in their bottom waters, which helped promote the excellent preservation of fossils in the Solnhofen limestones because these conditions were too harsh for most forms of life that usually decompose and scavenge carcasses that end up there. The fossils preserved in the deeper anoxic basins generally settled to the bottom after being washed in from the more oxygenated waters, reefs, algae-sponge mounds, or terrestrial settings in the surrounding area.

The most famous fossil species from the Solnhofen Limestone, of course, is the bird *Archaeopteryx*, but the array of species preserved in near-perfect condition is astounding: plants, sponges, jellyfish, corals, marine worms, brachiopods, bivalves, gastropods, squids and cuttlefish, ammonites and nautiloids; crustaceans similar to lobsters and shrimp, barnacles; horseshoe crabs, relatives of spiders; insects including mayflies, dragonflies, cockroaches, water skaters, crickets, bugs, cicadas, beetles, wasps, and flies; echinoderms including crinoids, starfish, sea urchins, and brittle stars; sea cucumbers; vertebrates including fish (chap. 3), sharks, rays, ratfish, many primitive and teleost ray-finned fish, coelacanths, turtles, ichthyosaurs, plesiosaurs, lizards, sphenodontians, crocodylomorphs, and the dinosaur *Compsognathus*. Also particularly well preserved in this deposit are the pterosaurs, the "wing lizards."

Wing Lizards

Pterosaurs are ornithodirans that, together with the broadly dinosaur-like lagerpetids, form the Pterosauromorpha. This latter group comprises reptiles closely related to but just outside the Dinosauromorpha (which includes the Dinosauriformes and Dinosauria itself). The pterosaur flying reptiles are not dinosaurs but they originated in the Triassic around the same time as dinosaurs and coexisted with them throughout the rest of the Mesozoic. The pterosaurs found preserved so well in the Solnhofen Limestone in the valley of the Altmühl include *Rhamphorhynchus* and *Pterodactylus* (fig. 11.1), two forms almost iconic for the pterosaurs in general, along with the geologically much younger *Pteranodon*. Ironically, my cousin and I had seen casts of some of these fossils in the Deutsches Museum but never realized we were so close to the field area. There are also specimens of some of these in the Goldfuss Museum at the University of Bonn, which I also saw, just years later when I could better appreciate what I was seeing.

Pterosaurs were reptiles with extremely thin, hollow bones; the limb elements in small animals such as those around during the Late Jurassic could be more than 5 mm (.2 in) in diameter yet have bone walls that were less than 1 mm (.04 in) thick, with the rest hollow. This hollowness allowed

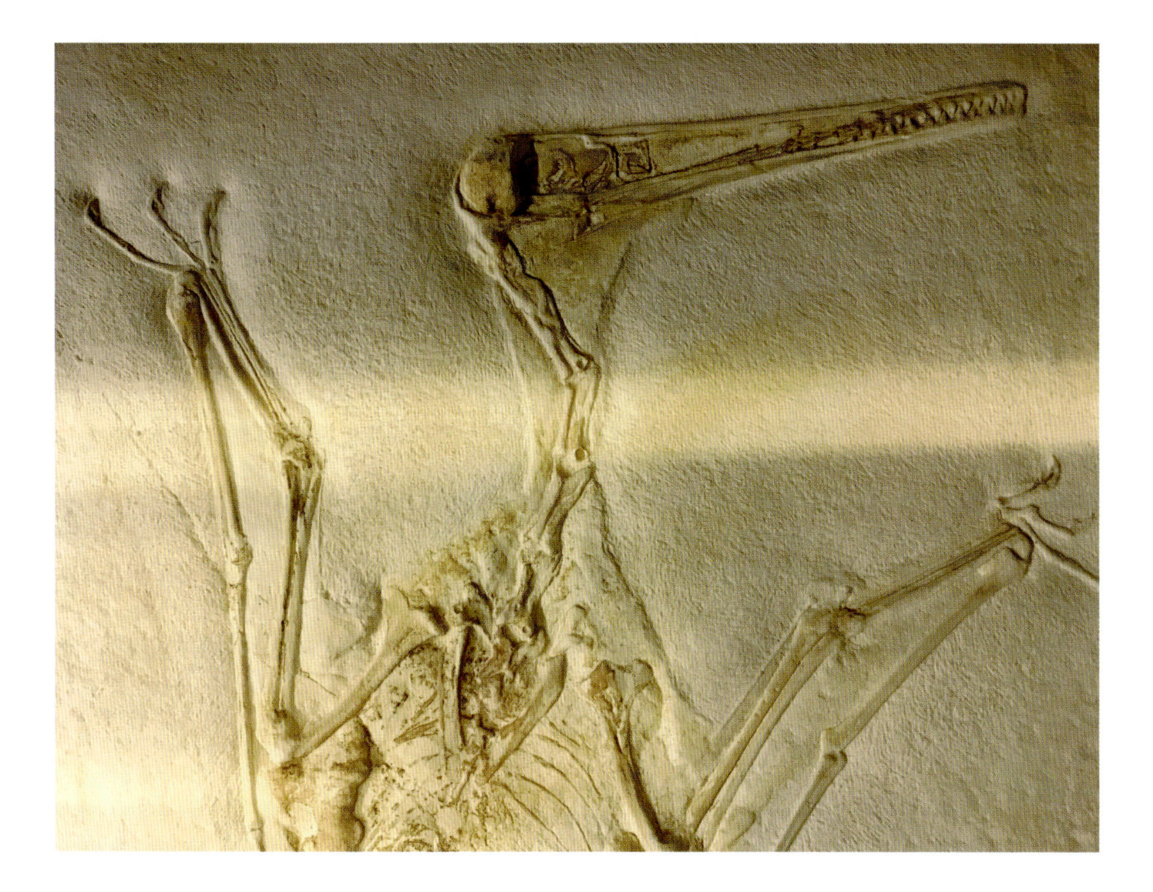

11.1 Front end structure of a pterosaur flying reptile, as demonstrated by *Pterodactylus antiquus* from the Upper Jurassic of Germany. Here, you can see details of the structure of the head, neck, anterior body, and wings.

pterosaurs to lighten their skeleton so that they could fly. Later forms took the lightening and hollowness to the extreme by keeping the bone walls around 1 mm thick even in much larger elements and strengthening the element by evolving very thin bony struts that crisscrossed the interior hollow of the bone—essentially buttressing the outer walls of the bone so the animal could get away with maximum hollowness.

The wings of pterosaurs are formed by the extreme elongation of their fourth digits, the "ring" fingers. The first three digits (equivalent to our thumb through middle finger) were short, clawed, grasping digits that the animal could use when not flying. The fifth digit had been lost. And in flight, the fourth digit was extended out as the leading edge of the wing, which was formed by skin that ran from the tip of the wing back to the hind leg. The wing consisted of thin muscle, skin, vascularization, and a system of leading-edge-to-trailing-edge stiffening rods known as aktinofibrils. The membrane of the wing filled in all the area between the hind leg and the body, upper arm, forearm, and hand (brachiopatagium). The leading edge of the wing was supplemented near the body by a skin membrane running from the shoulder to the wrist (propatagium), anchored there by a bone unique to pterosaurs called the pteroid bone. There also appear to have been short flaps of skin between the legs and tail (uropatagium).

The torsos and heads of pterosaurs may have been covered in hairlike, mostly single and unbranched filaments called pycnofibers, which would suggest that the animals were warm-blooded. These filaments have also been suggested to be decayed aktinofibrils, however, so there remains a bit of debate about pterosaur "fluff" covering. If separate material from aktinofibrils, these pycnofibers may be homologous to the nonflight feathers now known from so many feathered dinosaurs, as they seem to demonstrate some unique similarities. Perhaps less contentiously, the brains of pterosaurs appear to have been more complex than might otherwise be expected for animals of their size and grade, a condition that could be attributed to their need for neurological processing of complex information and movements associated with flight. It's not easy to fly, and as the first vertebrates to conquer the air, pterosaurs were pioneers in all the requirements of takeoff, landing, and midair maneuvering.

The exquisitely preserved pterosaurs of the Late Jurassic–age deposits in those stone quarries around Solnhofen and Eichstätt include seven different genera, but some of the best examples are *Aurorazhdarcho* (fig. 11.2A), *Rhamphorhynchus*, *Pterodactylus* (fig. 11.1), and *Ctenochasma*. *Rhamphorhynchus*, along with *Scaphognathus*, represents one of the major traditional groupings of pterosaurs, characterized by relatively small skulls with long, pointed, and somewhat forward-slanting teeth within the jaws, relatively short wingspans, and long tails with a diamond-shaped membrane near the tip. This broad, informal group is known as the "rhamphorhynchoids," and these were among some of the first pterosaurs known from the Triassic period. In some classifications, the broader group has been more precisely defined and split into a number of families. Rhamphorhynchids, one of those more restricted families, and the one retaining a name closest to the broader original one, are known mainly from Asia and Europe, and they may be present in the Morrison Formation, although the material from that formation is rather incomplete.

Pterodactyloids, on the other hand, exemplified by *Pterodactylus* ("wing finger") from the Solnhofen Limestone, had generally longer skulls with shorter but still pointed teeth (later forms lost their teeth entirely), relatively longer wingspans, and short tails. *Ctenochasma* was a specialized pterodactyloid with up to several hundred long, thin teeth in the jaws that pointed out from the mouth, forming a comblike structure that helped capture small prey items. It might have lived in a manner similar to modern nocturnal shorebirds. The Early Cretaceous–age *Anhanguera* from Brazil and Morocco was up to 4.5 m (15 ft) in its fully extended wingspan (fig. 11.2*B–D*). *Pterodaustro*, from the Cretaceous of Argentina, had incredibly slender, tapering, and upward-curving upper and lower jaws, with hundreds of especially long, narrow-diameter, closely packed teeth in the lower jaw that collectively formed a straining apparatus with which the pterosaur would filter-feed small food items out of water or sediment. Several probable ctenochasmatid coprolites (fossilized feces) from the Upper Jurassic of Poland indicate that these filter-feeding pterosaurs ingested shells of many

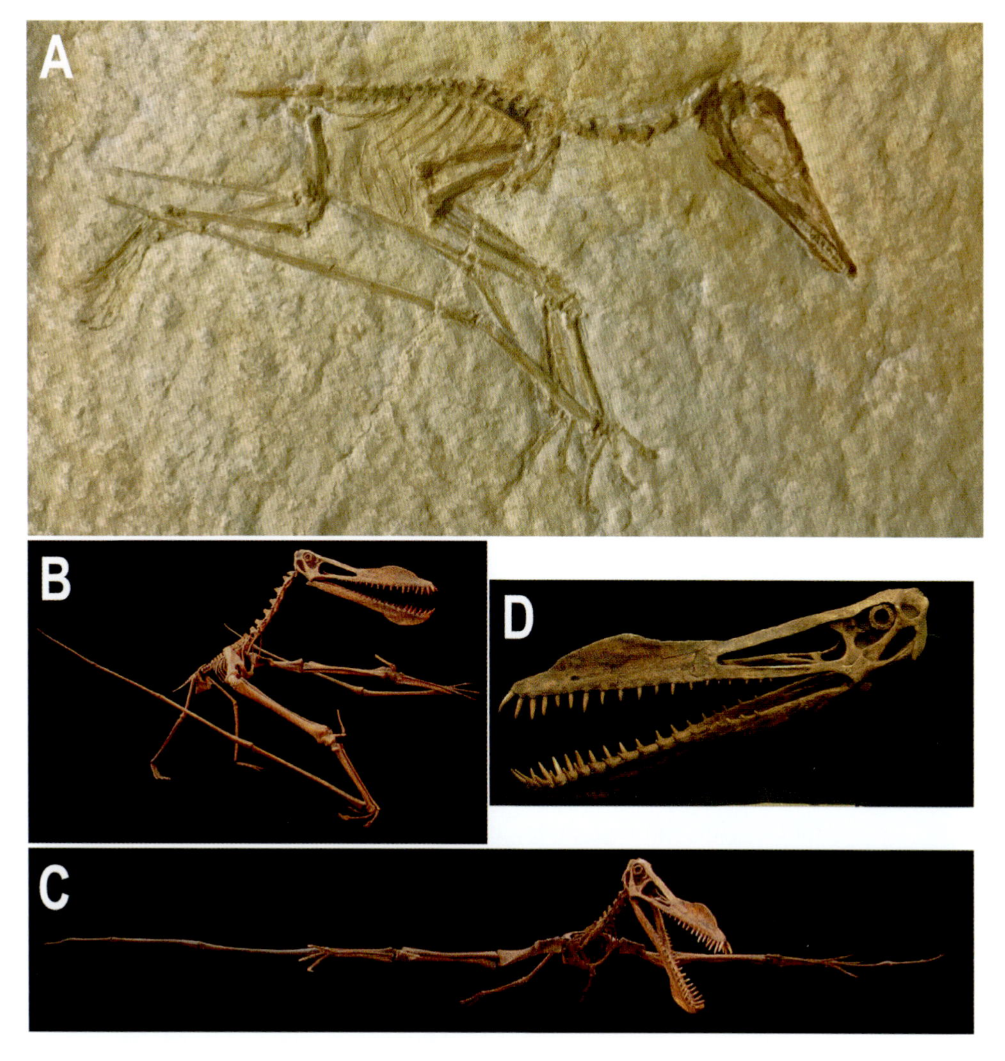

11.2 More pterosaurs. (*A*) The Late Jurassic ctenochasmatoid *Aurorazhdarcho* from Germany. (*B*) The large Early Cretaceous pterydactyloid pterosaur *Anhanguera*, from South America and northern Africa, in walking pose. (*C*) Same in flight pose. (*D*) Same with a close-up of the toothed skull. Note sclerotic rings near the back of the skull in *D* indicating the eye position. Total wingspan up to 4.5 m (15 ft). *Photos B–D courtesy of Rob Gaston/Gaston Design Inc.*

microscopic, foraminiferan single-celled eukaryotes, along with ostracod arthropods, bivalve mollusks, and possible polychaete annelid worms.

Another relative of *Ctenochasma* was *Cycnorhamphus* from the Late Jurassic of France and Germany. This latter pterosaur had relatively robust teeth that were restricted to the front ends of the upper and lower jaws, jaws that were also curved inward toward each other at those tips. The teeth seem to have been evolved to function similarly to ancient, biological nutcrackers, perhaps breaking open hard food items such as bivalve shells or other objects—and the jaws functioning like the business end of a pair of chisel-tip pliers. One thing ancient vertebrates can't be accused of is a lack of ecological disparity.

The tapejarid pterodactyloids from Brazil, Africa, Europe, and China were toothless forms with short skulls but tall, fan-shaped head crests that may well have been colorful in life as well. They may have been herbivorous to omnivorous. *Tupandactylus*, from the Lower Cretaceous Crato Formation of Brazil, may have one of the more unusual skulls of the Mesozoic, with a ventral flange on the dentaries, a bony "snout fan" (I made the term up) over the antorbital fenestra, small eyes, and a long, bony prong sticking out the back of the skull. Most unusual, there is a thin, curving rod of bone extending up and back from the "snout fan" to nearly over the back of the posteriorly directed bony prong. These structures suggest that a huge fan of skin existed between the curving rod and posterior prong. This entire structure is nearly as large as the rest of the body.

In 2001, my crew and I opened a quarry about a mile south of the Museum of Western Colorado's Dinosaur Journey Museum in Fruita, Colorado, because eight years earlier, some vertebrae of a juvenile *Camarasaurus* sauropod had been found weathering out of a hillside not far from the Fruita Paleo Area. We got one more vertebra of that animal but not much more, but more interesting, we found a tibia of a juvenile *Stegosaurus*, crocodile and turtle, theropod teeth, and a sphenodontid jaw. But most rare for the Morrison Formation, in 2002, I removed a piece of sandstone from this quarry while trying to get around the stegosaur tibia and found myself looking at two cross sections of a very hollow limb bone. We collected both of those pieces and several other nearby blocks that seemed to show some bone in them, and by the time my friend and field assistant, a bundle of energy and determination named Vaia Barkas, had prepared the blocks in our paleo lab, we had two or three wing elements of a pterodactyloid pterosaur. These were a radius, a wing phalanx, and another wing element not well preserved enough to identify.

Pterosaurs are even more rare in the Upper Jurassic Morrison Formation than the mammals. This may be in part due to their preferring areas near water, with the Morrison mostly representing floodplain, or it may be simply that the Morrison's more terrestrial environment is rougher on delicate bones than marine or lake environments in which they would get buried quickly in sediments. But either way, this rare occurrence near Fruita was one of only a handful of pterosaurs from this famous dinosaur formation. At first, we thought these small pterodactyloid wing bones we found might belong to the form *Mesadactylus*, also known from the Dry Mesa Quarry in western Colorado. But it turns out the Dry Mesa material might actually represent a chimera of several different types of pterosaurs (including a possible relative of *Ctenochasma*, in fact), so we probably won't be able to tell what the Fruita material is just yet.

Pterosaurs thrived in many environments throughout the Mesozoic, but many appear to have lived along the shores of seas and lakes and to have been fish-eating machines similar to modern birds like pelicans and others. They eventually evolved into the giant, high aspect ratio (in the wings),[1] toothless specialists we know in the famous forms such as the skull-crested *Pteranodon*, which has been found in deposits of the Late Cretaceous Western Interior Seaway of the Great Plains in North America, formations such as the Niobrara and the Pierre Shale. One summer, I visited my friends Bruce Schumacher and Dan Varner down in Edgemont, South Dakota; I say "down" because during the summers in those years, I was mostly working in the northern Black Hills and sometimes staying in Rapid City on off days. Bruce was a PhD candidate at the time at the South Dakota School of Mines, now with the Forest Service, and I had just recently finished my master's degree. Dan was a paleoartist who specialized in marine reptiles and other vertebrates of the Western Interior Seaway, a master of capturing underwater lighting (still not topped, in my opinion). Dan worked in the summers in what had once been the muds of the seaway, collecting all manner of marine vertebrates for the School of Mines. But the time I ventured out with them into the Pierre Shale on the prairie south of the Black Hills, on a break from my usual Jurassic routine and just for fun, we didn't see any marine reptiles. What we found in the dark mudstone of the Pierre were heavily gypsum-coated remains of large fish and an isolated bone of a large pterosaur, probably *Pteranodon*. *Pteranodon* was one of the large, high-aspect-ratio-winged and toothless pterosaurs that soared over the shallow seaways of the Cretaceous searching for fish to eat, and it had a relatively large head compared to some of those earlier pterosaurs of the Jurassic and Triassic. And then there was *Quetzalcoatlus*, a giant pterosaur of the Late Cretaceous of Texas with a wingspan of up to 10 m (33 ft). This giant animal had bones the size of those of many large dinosaurs and is, in terms of "press," the *T. rex* of pterosaurs, even though some other species were also fairly large and certainly stranger. *Nyctosaurus*, from the Niobrara Formation, for example, had an incredibly tall, bifurcated head-crest "antler" at least three times longer than the main part of the skull itself. The purpose or function of such a structure is not yet obvious to us.

Pterosaurs originated in the Late Triassic, a common theme here. Their diversity seems to have been highest in the Late Jurassic to the Early Cretaceous and is a little lower in the latest Cretaceous. Among the first-known forms in the Triassic are the dimorphodontids, animals with relatively short wings, long tails, and short, boxy heads with often several different types of teeth in the skull. These pterosaurs are known mainly from Europe, with a few forms known from South America and Mexico. "Campylognathoidid" pterosaurs are known only from Europe and Greenland in the Late Triassic and Early Jurassic and have long tails similar to *Rhamphorhynchus*, but they tend to have many closely packed and differentiated teeth, including enlarged caniniforms and many smaller multicusped cheek teeth that are superficially similar to those of triconodont mammals (three main cusps in a line, tallest in the center). These teeth even demonstrate wear facets, indicating that they chewed their food, unlike any other type of pterosaur.

Anurognathid pterosaurs were small, "Muppet-faced" (as paleontologist Mark Witton once put it), and appear to have specialized in catching insects in midair—and, also similar to today's bats, they may have swarmed the air in large clouds. Indeed, the skulls of these pterosaurs are unusually short, have very large orbital openings, possibly indicating specialization for nighttime hunting, and are rounded, without the elongate snouts of most other pterosaurs. And they are small—*Anurognathus* from Solnhofen would have been only a little over 10 cm (4 in) from snout to end of legs while flying. Members of the family are known from Asia and Europe and may also be present in the Morrison Formation in North America.

Ornithocheirids are among the most abundant and widespread pterosaurs, with specimens known from all continents except Antarctica. These pterosaurs were long-snouted with short teeth and short tails like pterodactyloids, and they had wide wingspans (up to 6 m or nearly 20 ft), with high aspect ratios.

The pterosaur *Darwinopterus* was recently described from the early Late Jurassic of China and appears to bridge a bit of the morphological gap between the traditional "rhamphorhynchoid" and "pterodactyloid" pterosaur groups. The skull and neck are pterodactyloid-like, while the tail is elongate and rhamphorhynchoid-like. One specimen of *Darwinopterus* was preserved with an unlaid egg associated with it, demonstrating the size and form of at least one genus's eggs.

More recent finds of pterosaur eggs in China preserved embryos inside. These specimens demonstrated that young pterosaurs may have been flightless for a period of time after hatching, similar to young birds. The geological context of the site and taphonomy of the egg deposit suggested that some pterosaurs nested in colonies to which they returned year after year.

Wing Tracks

In North America, the Morrison Formation's outcrops extend from southern Utah into the northeastern corner of Arizona, where they occur mostly in the Navajo Nation east of Kayenta. I once drove to a locality within hundreds of meters of the Utah–Arizona border, a site known as Boundary Butte, which had produced a few sauropod tracks and tracks of probable dryosaurid ornithopod dinosaurs in the Morrison Formation. I made a wrong turn before finding the locality and while turning around managed to get stuck in the sand miles from anywhere. This led to a long hike and a ride in the back of a pickup with some young men headed to town in Arizona—and then to many hours waiting for and then leading the tow truck in to unstick me. I eventually caught up with the rest of our field crew at Lake Powell that night, as planned; it was just at around midnight, not as planned. But in this same area of northeastern Arizona in the Salt Wash Member of the Morrison were found what proved to be the first recognized trackways of pterosaurs in the world. Described in 1957 by geologist William Lee Stokes, these hind footprints and accompanying handprints (named *Pteraichnus*) showed that pterosaurs walked on all fours with the first three fingers of the hand pointing forward and out and with the wing (the ring finger, again) folded back past the elbow.

These tracks have since shown up elsewhere in the Morrison and in the underlying Sundance Formation, including sites mostly in the lower and upper parts of the formations, respectively, as well as in other countries such as France and Korea. These trackways show a variety of walking styles among pterosaurs, varying somewhat in the gauge and spread of the hands, but all share the same general arrangement of the foot and hand digit impressions.

Pioneering Vertebrate Flight

The wings of pterosaurs are rather different from those of other flying vertebrates. Those of birds were formed out of flight feathers and an osteologically fused hand, while in bats, the fingers of the hand spread out within a wing membrane of skin. The pterosaurs lengthened their "ring finger" (fourth digit of the hand), lost the fifth digit, and retained the short, clawed first three digits. Just to review, the wing membrane (brachiopatagium) stretched from the tip of the wing finger to the back legs and up along the body and the trailing edge of the arm; this membrane was supported by very thin rods known as aktinofibrils, which radiated out from the bones of the leading edge of the wing toward the trailing edge and tip. The wing leading edge in front of the arm bones (humerus, radius, ulna) was formed by a membrane known as the propatagium, which anchored to the shoulder proximally and the medially pointed pteroid bone at the wrist. Finally, a membrane filling in the space between the hind legs, and sometimes incorporating the base of the tail, was called the uropatagium.

Pterosaurs developed this type of wing in the Triassic and thus were the first vertebrates to conquer the skies. Birds followed with their feather-and-stumpy-hand wing shortly after, at least by the Late Jurassic. And then bats evolved a wing with an arm and giant, splayed hand with skin between the fingers early in the Cenozoic (Early Eocene). Birds developed a range of feeding ecologies, whereas bat species are a bit more specialized in catching insects in flight or in feeding on nectar or fruits, blood, or even other small vertebrates. The development of echolocation in bats helped facilitate the insect-catching specialty in many species. Pterosaurs appeared to have been fish-eating specialists when they were first being discovered, and many species are, but more and more specialists have been found, and the ecological diversity of the group is now much greater than we realized.

The evolution of flight in vertebrates demonstrates how many times nature can achieve the same results when the need arises. Vertebrates had been on land for some time by the Triassic, when the pterosaurs took to the friendly skies, and it took relatively little geological time for birds to follow suit. Bats came along to join the birds about ten million years after pterosaurs disappeared at the end of the Cretaceous. Multiple evolutions of the same (or similar) ecologies in different groups of animals are quite common. And among animals generally, insects beat vertebrates to flying by tens of millions of years. Vertebrate gliding has evolved over the eons multiple times in mammals and reptiles. Even exocoetid fish can glide through the air for tens of meters (hundreds of feet)—and steer themselves while doing it! (And this capability has been around since the Triassic, in some forms, and since the Cretaceous, in exocoetids.) And other structures and capabilities of animals appeared independently more than once. Eyes have evolved at least 40 separate times in different animal groups! So the development of flight, complicated as it may be, is a "Mount Improbable" (to borrow a term from Richard Dawkins) that clearly is able to be summited by species when the opportunity and need arises—perhaps just with a little evolutionary "determination."

Note

1. Aspect ratio is essentially the length to width of a wing (length here being out from the fuselage or body). High aspect ratios are those of wings that are long for their width. Because most wings taper along their length, the aspect ratio can also be taken as the length squared divided by the wing area. A high-aspect-ratio bird wing might be that of a gull or albatross, and a lower-aspect-ratio bird wing would be a pigeon.

Feathered Dinosaurian Friends

Birds

12

IN THE OTHER DIRECTION from central Marienplatz in Munich—to the northwest not far from the river and the Deutsches Museum where my cousin and I spent so much time exploring—is the Palaeontological Museum of Munich, yet another paleontological treasure that I missed out on in those youthful wanderings decades ago. Housed in this museum (but not available during our visit to the city because it hadn't been found yet!) is the "Munich specimen" of the first bird, *Archaeopteryx*, one of the more complete specimens found. In a weird twist, this is actually the first (of only two) *Archaeopteryx* specimens I've seen in person, but I saw it in a temporary exhibit at the Field Museum of Natural History in Chicago in 1997 rather than returning to Munich. The second one I saw was the "Thermopolis specimen" just five years ago (fig. 12.1*B*). Both stopped me midstride and elicited reverent, speechless staring.

Archaeopteryx started out as a single-feather specimen described in 1861, but it was soon supplemented by a partial skeleton in 1861 and a nearly complete skeleton in 1874. The "Berlin specimen" of *Archaeopteryx* is one of the finest vertebrate fossil skeletons ever found (fig. 12.1A). Laid out on a slab of limestone about the size of a large pizza box, the bird is about as big as a large pigeon. Wings splayed out to either side with arms partially folded at the elbows and wrists, the fossil includes the imprints of the flight feathers of the wings in impressive detail for an animal that was buried in shallow marine mud in what is now Germany 150 million years ago during the Late Jurassic. The legs are preserved in right lateral view in what has the appearance of a "jumping" pose, the lower leg and foot noticeably longer than the femur of the upper leg. The vertebrae of the back have numerous ribs in place or nearly in place, and the neck curves up and back to the delicately preserved skull with sclerotic rings intact in the orbits and tiny teeth near the fronts of the jaws (fig. 12.2A). Feather imprints are also preserved on the lower right leg and on either side of the elongate tail (fig. 12.2*B* and 12.2C). The pubis bone of the pelvis is pointed down and back as in modern birds and dromaeosaurid dinosaurs (fig. 12.2*D*). Other specimens of *Archaeopteryx* confirm the presence of relatively long feathers on the lower legs.

Archaeopteryx was able to fly, and it was a bird, an avian dinosaur. It was probably not as powerful a flier as modern birds or even later Mesozoic ones. However, it was, so far as we can tell now, a pioneer. We've found no convincing evidence of birds taking to the skies earlier than this.

Recent work appears to show that at least some of the feathers of *Archaeopteryx* preserve fossil feather material and not just impressions. Analysis of this material in the original, isolated feather specimen of the bird showed

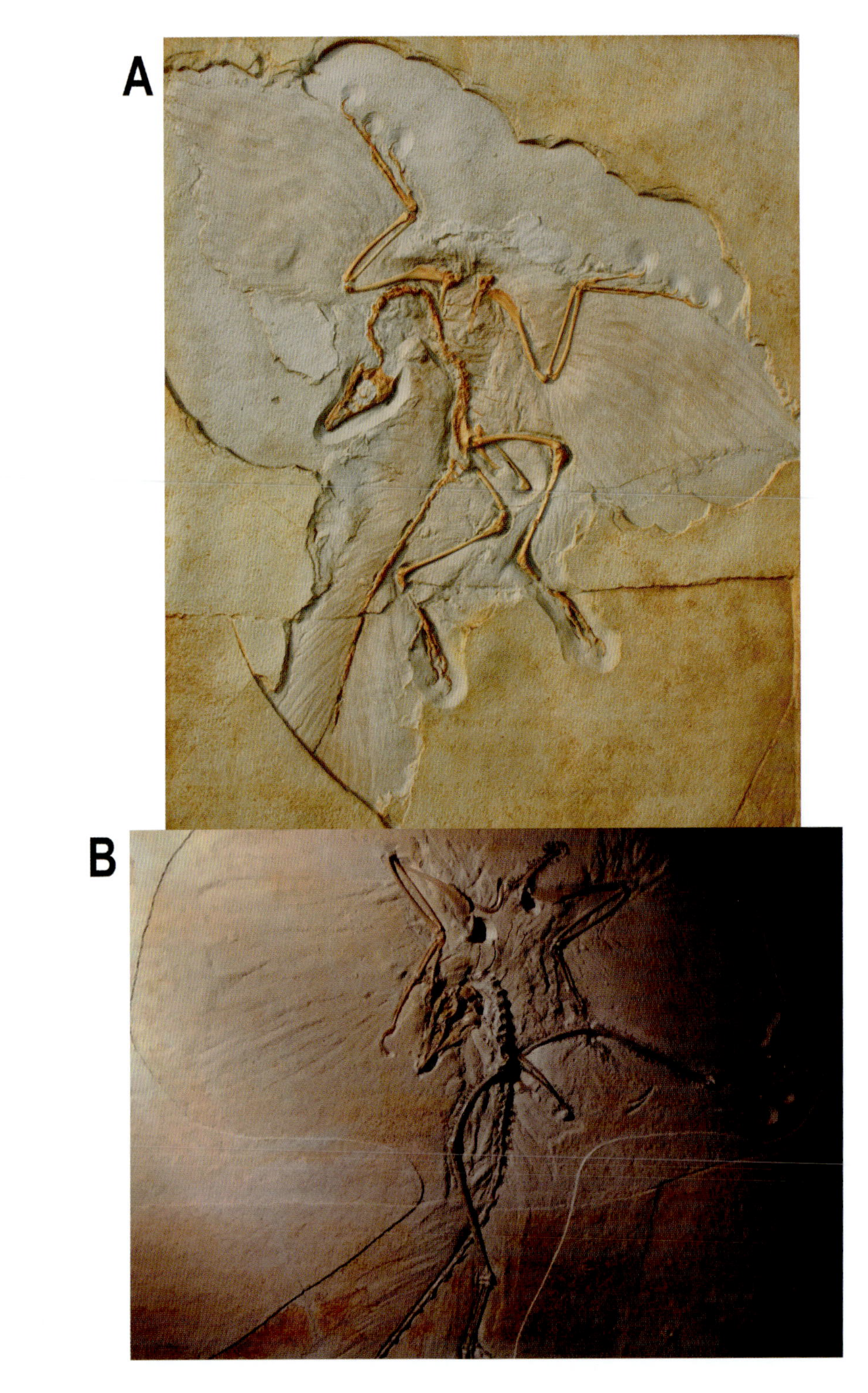

12.1 The Late Jurassic bird *Archaeopteryx* from the Solnhofen Formation of Germany. (*A*) Cast of Berlin specimen. (*B*) Thermopolis specimen as displayed at the Wyoming Dinosaur Center. *Photos by author.*

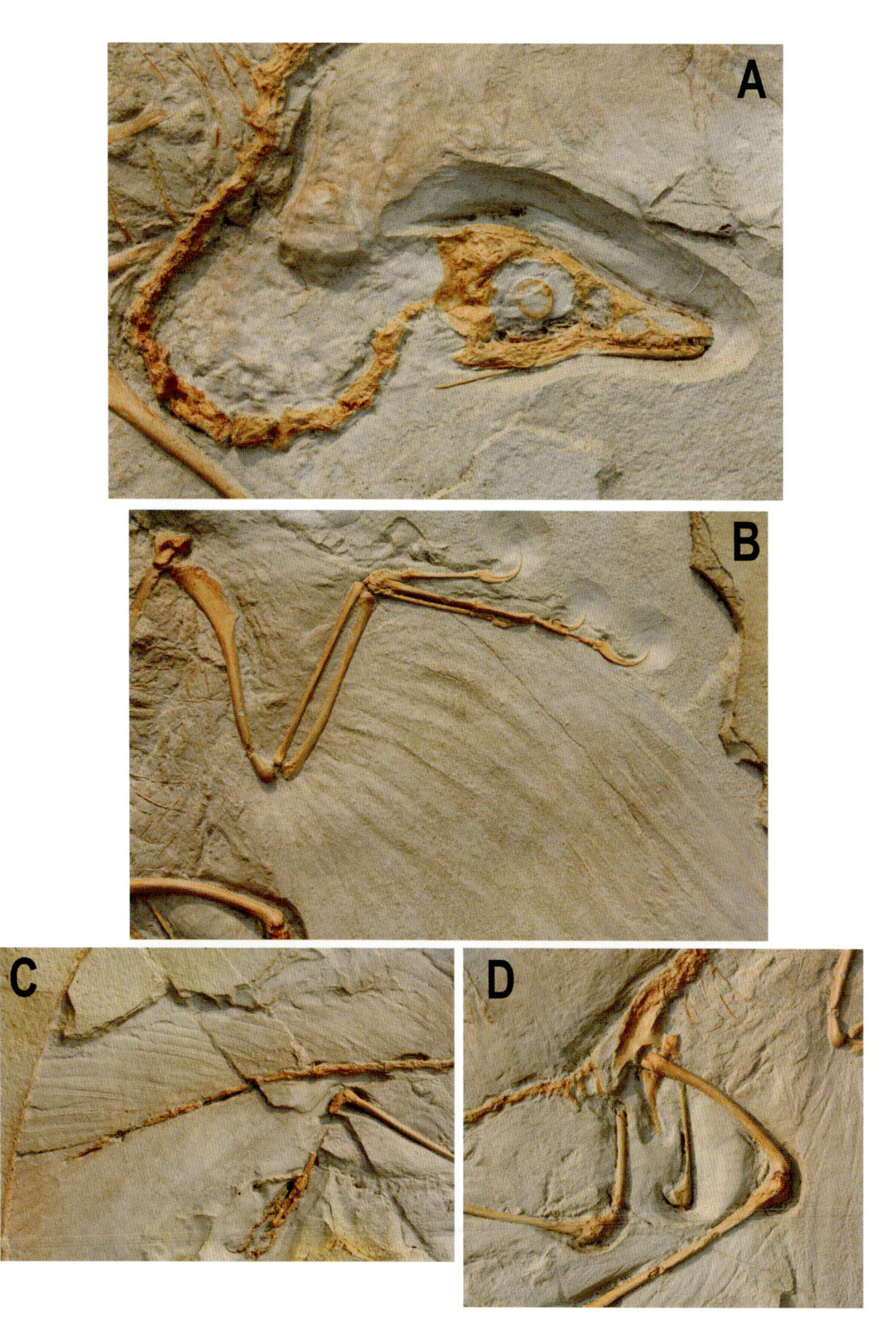

12.2 Details of the Late Jurassic bird *Archaeopteryx*. (*A*) Skull and neck vertebrae, featuring large orbit and teeth in jaws. (*B*) Structure of the wing and feathers. (*C*) Tail and feathers. (*D*) Hind limbs and pelvis, featuring the back-turned pubis bone. *Cast specimen, photos by author.*

that probable pigment coloring melanosomes are preserved, and these suggest that this particular feather was black. There has been some debate as to whether these microscopic structures were melanosomes or microbes, but that controversy has been resolved in favor of the former through the study of many additional fossils and modern feathers. Furthermore, studies of other animals, including feathered nonavian dinosaurs close to troodontids and dromaeosaurids, have suggested that *Anchiornis*, for example, an early Late Jurassic form from China that predates *Archaeopteryx*, likely had gray body feathers with black and white forearm feathers and a rusty reddish feather crest on the back of the head. The Early Cretaceous bird *Confuciusornis* appears to have had feathers of reddish or brown, gray, and black. Some birds and nonavian dinosaurs probably had feathers that were at least partially iridescent, as well as white feathers.

Avian dinosaurs today are everywhere. More than 10,000 bird species are spread across nearly all environments (and skies) and in forms ranging from sparrows, finches, and eagles to hummingbirds, penguins, and ostriches. The dinosaurs of today occupy just about any space you can find yourself in, from the opening of an underground cave to the vast plains of the open ocean. Their rise during the middle part of the Mesozoic led to a Cretaceous period that, globally, was likely surprisingly bird-rich, at least as far as we can tell from a few certain deposits with particularly good preservation that capture more of their respective ecosystems than most.[1]

Archaeopterygiids and Origins

There are deep pits in the Yixian and Jiufotang Formations in Liaoning, Hebei, and Inner Mongolia provinces, China, that contain thin-bedded lake deposits of Early Cretaceous age, among the sources of the famous Jehol Biota. This is one of the places that, starting in the mid-1990s, began to blur what we had previously thought of as the clear distinction between theropod dinosaurs on one hand, and birds on the other. Feathered dinosaurs and beast-companion birds side by side in the same ecosystem and of a diversity we'd never dreamed of, as we will soon see. Feathered theropods such as *Sinosauropteryx*, *Microraptor*, *Caudipteryx*, *Protarchaeopteryx*, *Beipiaosaurus*, *Sinornithosaurus*, and an ever-growing list of new species demonstrate that feathers covered most parts of the body in a variety of groups: troodontids, dromaeosaurs, oviraptorosaurs, and therizinosaurs. The bird fauna of the Jehol Biota is an extraordinary addition to this already remarkable diversity of feathered theropods. As implied above, the feathers of birds were nothing new, having been inherited from the nonavian theropods. These feathers likely started out as insulation or for displaying color and simply became modified into flight feathers in birds and their closest dinosaurian relatives among theropods. The origins of feathers of any kind among theropod dinosaurs appear to go back quite deeply into the theropod family tree.[2]

Birds, starting with *Archaeopteryx*, arose from the group of theropod dinosaurs that includes dromaeosaurs (think *Velociraptor*), scansoriopterygids, and troodontids. These groups and *Archaeopteryx* itself were present by the Late Jurassic. Scansoriopterygids were very small, feathered relatives

12.3 The Early Cretaceous bird *Jeholornis* from China. *Photo courtesy of Stephanie Abramowicz and Luis Chiappe, and from Chiappe and Meng, 2016 (*Birds of Stone*).*

of dromaeosaurs and troodontids that may have been semiarboreal gliders.[3] From this small-sized theropod starting point among dromaeosaurs, troodontids, and scansoriopterygids (animals with long, bony tails, toothed jaws, grasping, clawed hands, and an otherwise still "dinosaurian" body plan) developed rather quickly what we recognize as a birdlike design: short, bony tails consisting more of feathers than bones, more compact bodies, and toothlessness (in some lineages), among some others. But not all bird lineages went more modern birdlike—it appears a few retained *Archaeopteryx*-like proportions and long tails for quite some time. In Madagascar, the possible bird *Rahonavis* was a bit of a morphological throwback to earlier birds during its time in the Late Cretaceous, around the same time *Pentaceratops*, *Parasaurolophus*, and tyrannosaurs were lurking and lounging about the deltas of the Rocky Mountain region in North America, some 75 million years after *Archaeopteryx*. Temporally linking these two archaic bird designs, *Jeholornis*, from the Early Cretaceous of China (fig. 12.3), is another toothed and long-tailed *Archaeopteryx*-like bird from the Jiufotang Formation. The stomach contents of *Jeholornis* suggest it ate many seeds and fruits. Apparently, there was a place for this original, long-tailed, *Archaeopteryx*-like bird design in the ecosystems of the second half of the Mesozoic, not just the Late Jurassic, and the headlong rush toward "modern birdness" in some bird lineages was not necessary for all.

Confucius Birds

Confuciusornithids lived alongside *Jeholornis*-type forms in the Jehol Biota and were the next evolutionary grade of birds. These were toothless, short-tailed, and crow-sized birds such as *Confuciusornis* (fig. 12.4A), preserved in literally hundreds of partial to nearly complete skeletons in the Yixian and Jiufotang Formations. They are so relatively abundant, as vertebrate fossils go, I once saw one for (questionably legal) sale in a shop on Larimer Street in Denver, of all places—yours for a mere $20,000! Thankfully, one can also have full-time, legal access to examine a good cast of a specimen for about 0.3% of that cost. In some *Confuciusornis* specimens, likely representing sexual dimorphism, there are two extralong feathers extending far out from the back of the tail, longer than the main body itself. There is a boomerang-shaped furcula, the fused "collarbones" known in living birds like the turkey as the wishbone, another element present in similar form in many theropod dinosaurs. Although the skull has a beak rather than teeth (fig. 12.4*D*), the neck and body of *Confuciusornis* are fairly similar in proportion to *Archaeopteryx*, even though the articulations of the neck vertebrae are slightly closer in form to modern birds. The fingers of the hand were still fully separate and had large claws much like *Archaeopteryx* (fig. 12.4C). The sacrum is more elongate and fused more like today's birds' synsacrum than older forms. The pubis and ischium were more birdlike, and the tibia, astragalus, and metatarsals were approaching a more modern birdlike form. Most notably, *Confusciusornis* had a pygostyle of fused tail vertebrae, broadly similar to those of modern birds (fig. 12.4*B*), which supported a short, diamond-shaped fantail of feathers that likely functioned in midflight steering assistance.

The toothlessness of confuciusornithids is one case of at least six independent losses of teeth (and switches to an exclusive beak structure) among theropods. Some ornithomimids and oviraptorosaurs switched away from teeth, and then among birds, confuciusornithids were one of three groups

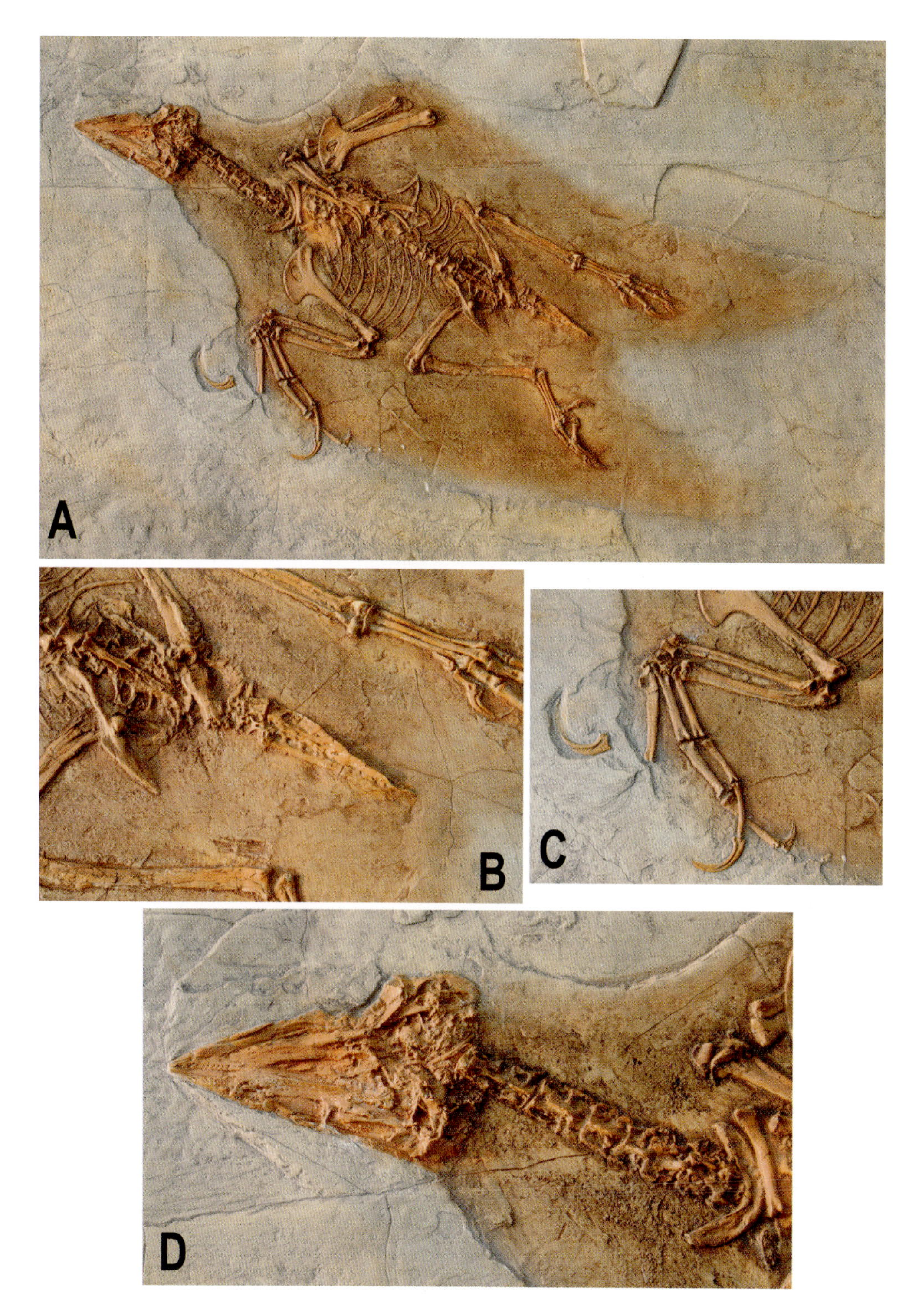

12.4 The Early Cretaceous bird *Confuciusornis* from China. (*A*) Full specimen. (*B*) Detail of pelvis and pygostyle. (*C*) Detail of wing bones. (*D*) Detail of skull and neck vertebrae. *Cast specimen, photos by author.*

in addition to modern birds (Neornithines) to do the same. Teeth were retained in the closest relatives of living birds, including *Ichthyornis*.

Opposite Birds

Another step more derived among the Cretaceous birds, and a sister group to modern avians, were the Enantiornithes. This is a rapidly growing group of about 80 genera, and as a group, they have been reported from all continents except Antarctica, with a particular diversity known from China. They represent the first real diversification event in the history of birds, they are the most diverse group of Mesozoic birds, and they diversified before *T. rex* or *Triceratops* were even around.

Enantiornithines (fig. 12.5) have a number of features that are closer to the conditions in modern birds than the Mesozoic forms we've seen so far, but they still retain a few more primitive aspects of their morphology. Most of these birds had teeth, although several species have lost them partially or completely. In most enantiornithines, the neck is somewhat longer and the body slightly shorter than in previous birds; the coracoid bone (in the shoulder) is longer, which helps amplify the flight stroke of the wing; the furcula, or wishbone, has become more V shaped, which helps add a bit of spring to the recovery stroke of the wing; a wing tendon has developed a pulley action over the scapula, coracoid, and furcula in the shoulder, which also helps with the recovery stroke (the presence of this tendon is indicated by the triosseal canal at the junction of these three bones); the keel on the sternum is present and enlarged, which allows enlargement of the breast muscles powering the wing; the radius and ulna have lengthened relative to the humerus; the metacarpals have begun to fuse to the carpals; and the outer digit of the hand has been reduced and has lost its claw. All of these changes are relative to more primitive birds and are closer to the condition of birds we see today.

Equally as important, the enantiornithines developed a diversity of morphologies and ecologies reflective of their diversification event. Many were small, sparrow- to thrush-sized, which reflects increased flying capabilities and a diversity of ecologies. There were species that specialized in perching and possibly swimming and wading, from what we can tell just from the variety of foot designs. Some had wings with high aspect ratios, suggesting an ability to glide much of the time (think of an albatross among modern birds), and others with lower aspect ratios that were probably power fliers. Enantiornithines appear to demonstrate a range of reproductive strategies too, with some forms having been found with a large number of small eggs still in the body cavity; others employed an opposite strategy and had fewer, larger eggs. Variations in skull design and tooth arrangement indicate different feeding modes, including possible seed specialists, piscivorous species, and forms that may have specialized in probing the sediment for food items in a manner similar to modern shorebirds. Enantiornithines even appear to have been food for, of all things, ichthyosaurs and dromaeosaurs (*Microraptor*), at least in rare cases. And material has been found in areas that during

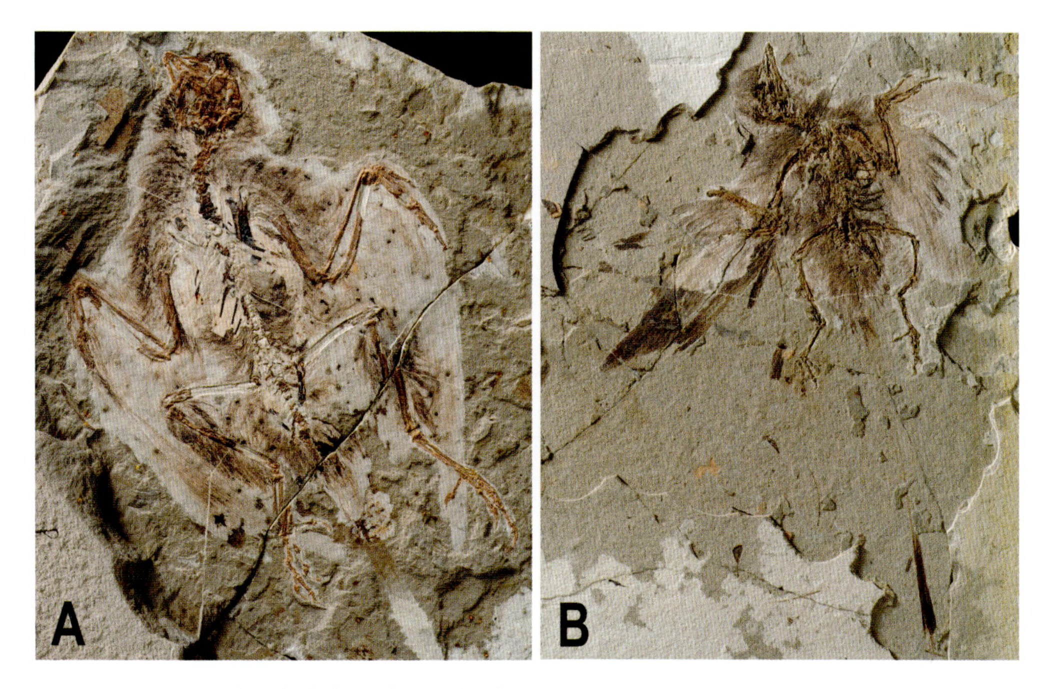

12.5 Some enantiornithine birds from China. (*A*) Indeterminate enantiornithine bird from the Early Cretaceous Jehol Biota. (*B*) The Early Cretaceous *Protopteryx. Both images courtesy of Stephanie Abramowicz and Luis Chiappe, and from Chiappe and Meng, 2016 (*Birds of Stone*).*

the Cretaceous would have been within the Antarctic Circle, suggesting that these birds already may have developed long-range migration. After all, why stick around for the months of darkness? On the other hand, there are ducks that overwinter in arctic and Antarctic regions today, so ancient species may also have lived at high latitudes full-time.

The first recognized enantiornithine, *Enantiornis*, was excavated in 1974 in the Lower Cretaceous of Argentina (along with bones of the sauropod dinosaur *Saltasaurus*) and was described in 1981. Also in 1974, *Gobipteryx*, a toothless form, was found in Upper Cretaceous rocks in Mongolia. Around the same time period, a crew from Occidental College and the Natural History Museum of Los Angeles County, led by William Morris (my undergrad adviser Don Prothero's predecessor at Oxy), found a form called *Alexornis* in Baja California, a sparrow-sized bird that had lived on the tropical coastline of what would become Mexico, during the Late Cretaceous. It took until 1983 for all three of these birds to be recognized essentially as enantiornithines, a group of interrelated birds, in a study by paleontologist Larry Martin. Not long after recognition of this group, more members began coming fast and furious. *Iberomesornis* was found in the Las Hoyas lagerstätte of Spain. *Lenesornis* was found in the Late Cretaceous–age rocks of Uzbekistan. The Late Cretaceous of Argentina produced *Neuquenornis* while *Eoenantiornis* and *Cathayornis*, among others, came out of the

12.6 The Early Cretaceous bird *Longipteryx* from China. *Photo courtesy of Stephanie Abramowicz and Luis Chiappe, and from Chiappe and Meng, 2016 (*Birds of Stone*).*

Early Cretaceous of China. *Halimornis* was found in Late Cretaceous–age deposits in Alabama that, during those ancient times, would have been about 50 km (30 mi) offshore, suggesting that this bird was a coastal flier that ventured well out to sea now and then. *Gettyia*, from the Upper Cretaceous Two Medicine Formation of Montana and named by Jessie Atterholt, Howard Hutchison, and Jingmai O'Conner for the Denver Museum's Mike

Getty, appears to have had legs built for strength more than speed in their flexing action, something emphasized in predatory birds and those that are arboreal and sometimes perch on branches while feeding (e.g., parrots, chachalacas). The closely related *Mirarce*, from the Kaiparowits Formation of Utah, is the first enantiornithine found to have quill knobs on the ulna bone of the forearm. In modern birds, quill knobs help to strengthen the anchoring of the flight feathers to the forearm.[4]

The Early Cretaceous *Protopteryx*, also from China, is small (about the size of your hand) and among the most primitive of enantiornithines (fig. 12.5*B*). It has tufts of feathers covering the head and body and long flight feathers on the wings and a short set of tail feathers—except for a pair of extralong feathers, sticking out from the rest of the tail, that are about the same length as the rest of the body.

Longirostravis, another Early Cretaceous form from China, had an elongate, curved snout and tiny, conical teeth just at the tip, suggesting it may have engaged in a mud-probing feeding strategy. *Longipteryx* also had an elongate snout with the teeth at the tip (fig. 12.6), but the teeth were much larger and recurved, and the bird may have fed on fish. Yet the two birds may be closely related. *Eoalulavis* was the size of a goldfinch or kingfisher and was found in the Las Hoyas deposit in Spain. Its fossil preserved an alula, or winglet, three to five flight feathers attached to the first digit that, in modern birds, is moved up and forward to assist in increasing lift at low speeds, just like the leading-edge slats that are deployed on a modern airliner during takeoff or landing. Alulas were also found on *Protopteryx*.

Bird Tail Shapes

Enantiornithines are the sister group to a clade of birds known as the Ornithuromorpha ("bird tail shape"), the latter a group that includes modern birds (the Neornithes) plus a number of stem groups. Among these basal ornithuromorphs, getting closer to modern birds, was *Patagopteryx* from the Late Cretaceous of (surprise) Patagonia, Argentina. This was a chicken-sized animal found in the early 1980s, and it had reduced wings and relatively large hind legs. The pelvis was wide, perhaps indicating it laid large eggs, and the neck was long and probably capable of an S-shaped posture. Given its small wings, large legs, and wide pelvis, *Patagopteryx* is believed to have been secondarily flightless.

Other basal ornithuromorphs are *Yanornis* and *Yixianornis*, both from the Early Cretaceous of China, forms that have teeth, advanced shoulders for improved wing movement, two clawed fingers with just the beginnings of fusion in the hand, and enlarged keels on the sternum—*Yanornis*, however, has an elongate, triangular skull and appears to have been a fish eater. *Apsaravis*, from the Late Cretaceous of Mongolia, was a toothless, sharp-beaked ornithuromorph that had developed a modern-level innermost bone in the hand (metacarpal) that allowed improved function of the alula and automatic extension of the hand and wing when needed, a capability typical of today's birds. Some ornithuromorphs have been found with small stones in the body cavity, suggesting these birds had a gizzard. Other types of birds, like *Sapeornis* from Jehol, indicate that some had a predigestive crop in which food like seeds and fruits were stored.

Bird Tails

The pigeon-sized *Gansus* (Early Cretaceous of China) is another ornithuromorph but is considered a member of Ornithurae ("bird tail," a clade of more derived ornithuromorphs including Neornithes and a few stem forms such as *Hesperornis*). *Gansus* appears to have been ecologically similar to ducks in that it could fly and was aquatically adapted. It also likely possessed the ability to propel itself over the water's surface and maybe even dive underwater with its feet. Even its overall body proportions are broadly reminiscent of ducks. Another flying and aquatic form, *Tingmiatornis*, was found in the arctic of Canada and was only recently described from partial remains; it likely was capable of subsurface diving.

Among the other Ornithurae outside modern birds are, as mentioned, *Hesperornis*, from the Late Cretaceous Western Interior Seaway deposits of the Great Plains, and a few other birds also adapted to foot-propelled marine and riverine diving. *Hesperornis* has been found from western Kansas to arctic Canada and was a toothed, piscivorous bird with very reduced wings, long legs with large, probably lobed toes, and a long snout with a beak at its tip. The bird was up to 2 m (6 ft) long and swam the shallow marine stretches and rivers of the Cretaceous. O. C. Marsh's 1880 monograph on *Hesperornis* and other related Mesozoic birds drew from the sages of the US Congress mockery about the preposterousness of birds with teeth. This reaction contributed to a cut in funding to the United States Geological Survey during John Wesley Powell's tenure as its director—yet another of those oh-so-rare instances in which lawyer-politicians anointed themselves the experts in what was and wasn't solid science and saved us from ourselves. Luckily, the *Hesperornis* fossils remain, absurd teeth and all.

Other diving, marine birds of the Great Plains and Canada include the basal hesperornithine *Pasquiaornis* and the more derived *Baptornis*, which are smaller than *Hesperornis*. There appears to be some direct evidence, in *Baptornis*-associated coprolites, that hesperonithiforms ate fish, such as *Enchodus*. A possibly freshwater hesperornithiform was recognized recently in *Brodavis* (and likely the junior synonym of *Potamornis*), known from the Late Cretaceous of Canada and South Dakota, plus Mongolia. In recent years, hesperornithiform finds have taken off in the field and in museums, with forms reported from new areas such as the Late Cretaceous of Japan (*Chupkaornis*), the Late Cretaceous of Kazakhstan (*Asiahesperornis*/*Parahesperornis*), and familiar areas like the Niobrara Formation of Kansas (*Fumicollis*, *Parahesperornis*), highlighting a previously underappreciated diversity of these diving birds.

Seabirds like gulls and scissorbills have their Mesozoic approximate ecological equivalent in the derived *Ichthyornis*, a partially toothed bird close to modern Neornithes and known from Western Interior Seaway deposits of

North America, ranging from Alberta and Saskatchewan through Kansas, New Mexico, and Texas, and down to Alabama and California. The sternum keel and the fusion of the bones of the hand in the wing of *Ichthyornis* are the most modern grade of any of the birds we have encountered so far. Fossils of *Ichthyornis* suggest an animal about the size of pigeons to crows.

Old New Birds

The Late Cretaceous *Limenavis* is another advanced member of Ornithurae, known from Argentina. A number of Mesozoic fossils attributed to modern groups within Neornithes indicated that our current avifauna diversified at the end of the Age of (nonavian) Dinosaurs, but the fossils involved almost always proved to be single or isolated bones, which are nearly impossible to refer to known groups in most cases. Still, apparent representatives of Neornithes exist in Late Cretaceous–age rocks. Among these, *Vegavis*, from the Antarctic Peninsula, appears to be a close relative of today's ducks and shares with them not only some details of anatomy but also general body form and ecology. And *Teviornis*, from the Late Cretaceous of Mongolia, may be related to what are known as (get ready) the presbyornithid anseriforms, meaning that it also may be a relative of ducks (the anseriform part) but one that might have been a long-legged, wading, ecological equivalent of flamingos along lake shorelines (the presbyornithid part).[5] In any event, these two Cretaceous birds, *Vegavis* and *Teviornis*, suggest that the lineage that led to ducks, geese, and swans (the anseriforms) had appeared before the end of the reign of nonavian dinosaurs and that, by extension, their sister taxon the galliforms (leading to chickens, grouse, turkeys, and others) must also have been on the scene by the same time. The more primitive Palaeognathae should also have been present in some form at this time, if our understanding of modern bird phylogeny is correct, suggesting that ancestors of ostriches lurked somewhere in the Cretaceous world. Not necessarily immediate ancestors, but ancient relatives nonetheless.

Bird Tracks

Bones are not the only indication of birds' presence in Mesozoic rocks. They left footprints too, although, of course, one individual can leave thousands of them. In some ways, it is surprising we don't see even more tracks of Mesozoic birds, but there is no dearth of them either. Such tracks, of a decent variety too, have been reported from the western United States, Canada, Korea, China, Japan, Europe, South America, and Africa. These tracks can give us indications of paleoenvironments, trackmaker paleobiology, and taxa that might otherwise be missing from particular units.

Running between the Utah–Colorado border and the west side of the San Rafael Swell along Interstate 70 in Utah is a stretch of outcrop of the Cedar Mountain Formation, an Early Cretaceous–age unit that directly overlies the famous Morrison Formation of Dinosaur National Monument fame. The Cedar Mountain blends in with the Morrison in many places, and the contact can be difficult to distinguish, but it is some 5–10 million years younger, at its lowest reaches, than the top of the Morrison, and it contains a distinct fauna of dinosaurs that have come to light mostly in the past 30 years. Ironically, the lithologic similarity was noted early on in study of the Cedar Mountain, and one of the early characteristics used to distinguish the Cretaceous formation from the Morrison was, of all things, a lack of dinosaurs! There are now probably more types of dinosaurs known from the Cedar Mountain than from the Morrison! The Cedar Mountain is now famous for dinosaurs such as the dromaeosaurid *Utahraptor*, the polacanthid ankylosaur *Gastonia*, and the basal therizinosaur *Falcarius*. And there are numerous dinosaur faunas that turn over through time during the 19 million years of Cedar Mountain time. There is a whole host of other animals from the formation too, not just dinosaurs, but one thing that is missing is bird bones. But that doesn't mean there weren't birds around western North America during the Early Cretaceous. There were birds living along the lake shorelines along with some of the dinosaurs of the time, birds we know about because they left behind footprints on the sands of time.

Rob Gaston is an artist of the molding and casting of the bones of dinosaurs and the mounting of their reconstructed skeletons. A former furniture maker and explorer of the productive paleontological outcrops of western Colorado and eastern Utah, Rob reconstructs skeletons from all over the world in a sprawling property packed with shops and storage in Fruita, Colorado. Because of his previous time prospecting, Rob knows where a lot of good outcrop areas are. One day, a number of years ago, Rob and I headed to Moab, Utah, so he could show me some outcrops with numerous bones and teeth in the Chinle Formation not far from town. On the way, we had agreed to help Jim Kirkland, state paleontologist of Utah, jacket and remove some material from a site on a slope in the Cedar Mountain Formation north of Arches National Park. During a break in the jacketing of the material, Rob and I decided to run down to the car to grab more snacks or water, and we happened across blocks of sandstone with some three-toed tracks in them. I looked at one, and Rob wandered over to another.

"Nice theropod track here," I said.

"Birds over here," Rob commented rather matter-of-factly.

I jumped away from the theropod to the avian theropod tracks. Indeed, Rob's slab just a few feet away had more than 50 small tracks with widely splayed, short digits in tracks forming narrow, wandering, short-stepped, and pigeon-toed trackways. They looked very much like tracks of modern shorebirds milling around the sand on the shoreline.

Because these were on what seemed to be a route between Jim's quarry and the vehicles, we assumed they were known to the crew and didn't think much more about them on our way to the trucks and back. When we got back to the quarry, I casually mentioned to Jim, "Nice bird tracks down there."

"Nice what?" Jim asked.

Apparently, Jim and the others had not yet happened across the block Rob and I had seen, though until this point, we assumed they had. We had just taken a path a few feet from the barely established one taken by everyone else up to that point. It turned out there were more blocks with more tracks on them in the immediate area, more than 130 tracks in at least 43 trackways. Along with Jim, Martin Lockley, Joanna Wright, Lisa Buckley, and Don DeBlieux, we eventually determined that these tracks belonged to an ichnotaxon called *Aquatilavipes*, a fossil track type morphologically similar to tracks of modern shorebirds, although whatever the Cedar Mountain birds were—enantiornithines?—they were not likely part of a modern group. We now knew for sure that birds lived alongside Utah dinosaurs in Cedar Mountain Formation times, an interval approximately equivalent to the time of the Jehol Biota in China. They probably were frequenting lakeshores, possibly to feed.

Bird Lessons

So the feathery world of the Mesozoic was bursting at the seams with feathered dinosaurs, both avian and nonavian, the line becoming harder for us to distinguish with each new find from the dromaeosaurid-archaeopterygiid neighborhood of the theropod tree. We can learn several things from this menagerie of regal ancient birds and bird relatives: (1) feathers evolved well before flight; (2) thus, flight feathers were effectively an exaptation, a modification of an existing feature (in this case, insulation and pigment-carrying integumentary structures) for a new purpose (lift and eventually flight); (3) birds went through a number of stages of becoming more and more modern grade in their morphology, osteology, and flight ability, but these various forms seem to have co-occurred in time (and sometimes in the same deposits) during the second half of the Cretaceous especially; (4) birds of modern aspect, basal members of some of today's groups, appeared during the Late Cretaceous, setting the stage for a second, this-time-Cenozoic diversification of birds after nonavian dinosaurs became extinct; and (5) dinosaurs live on in the birds of the Age of Mammals, and today, these dinosaurs are possibly more diverse than they have ever been. They are, of course, around us almost all the time. We hear them in the squawk of the magpie, the coo of the mourning dove, the song of the sparrow, the earsplitting screech of large parrots, and the otherworldly "whoo" of an owl. And we see their skeletal anatomy frequently at mealtime. Duck or pheasant for special meals, Thanksgiving turkey (in October or November), the guilty pleasure of a bucket of fried chicken when it is fireworks time . . . avian dinosaurs were made for holidays! Our world is richer and certainly more colorful thanks to these lucky survivors from the Cretaceous.

Notes

1. Some fossil bird specialists, such as Alan Feduccia and the late Larry Martin, have maintained that bird origins probably lie among early Ornithodira rather than within derived Theropoda, but nearly all phylogenetic analyses have birds coming out next to troodontids, dromaeosaurs, and so on (i.e., well derived within Theropoda). A large majority of paleontologists accept birds as dinosaurs unless and until a better-supported phylogenetic hypothesis is presented.

2. Feather-like structures of some form appear to be present in some ornithischian dinosaurs as well.

3. The scansoriopterygids *Yi* and *Ambopteryx* are among the strangest of feathered theropods in that they appear to have had a full wing membrane and a divergent digit in the hand supporting the membrane. This bat-like arrangement was unknown in dinosaurs, avian or nonavian, before 2015.

4. The dromaeosaurid dinosaur *Velociraptor* was found to have quill knobs, indicating not that it flew but that it had feathers.

5. Presbyornithids were common waders around the Eocene lakes of North America about 15–20 million years after the Late Cretaceous ended.

Copycatted

Mammals and Synapsids

13

THERE MAY BE NO BETTER COMBINATION in Mesozoic paleontology than the Late Jurassic–age deposits of Portugal, paired as they are with the beaches on which they frequently occur, with dinosaurs on the cliffs above the sand and waves. And it doesn't hurt to have wine country nearby. Jurassic fossils, surfing, and wine all in one spot. I'm jealous of this arrangement. And some of the largest waves ever surfed were ridden off Nazaré, up the coast north of Lisbon.

Just a bit inland and northeast of that surf spot is the city of Leiria. The southeastern edge of Leiria boasts the coal mine of Guimarota, one of the most productive Late Jurassic fossil localities in the world. Decades ago, the coal of the Alcobaça Formation, under a small village on the outskirts of Leiria, was mined to supply power fuel to the cement plants of the city. Fossils were discovered in the coal in the late 1950s shortly before the mine closed, and the mined coal was inspected for several years in the early 1960s by paleontologists from Germany. By the 1970s, the mine had been closed for a few years and had flooded. When paleontologists went back in to mine the coal specifically for the fossils, they first had to pump out and shore up the tunnels. But they had hit one of the Mesozoic mammal fossil motherlodes of paleontological history.

The site is nearly within the edges of Lieria now, and the property, while still recognizable, is no longer a mine of any kind. It is now surrounded by apartments and businesses, including a racing auto parts store, a shoe store, and (of all things) a McDonald's. Nothing says "growth" like the appearance of the Golden Arches even near a shrine of paleontological history, but then, there is a KFC across from the Sphinx at Giza, so the infiltration is widespread. In its day, the Guimarota mine yielded nearly 8000 Late Jurassic mammal fossils, from isolated teeth to jaws to whole skulls and even two nearly complete skeletons.

The Guimarota mine was constructed in the 1930s to access two coal seams, separated by several meters of limestone, in the Upper Jurassic Alcobaça Formation. Although it has not been precisely dated within the Late Jurassic, all indications so far put the deposit in the early Kimmeridgian age, approximately equivalent to the lower to middle parts of the Morrison Formation of the western United States. At the time, Europe was a collection of islands in a tropical sea situated between the Tethys Sea to the south and east and the early Atlantic Ocean to the west. The Guimarota deposit appears to have been a swampy coastal deposit on the edge of one of those islands with the sea to the south, as it contains a mix of layers with some indicating terrestrial deposition and others indicating shallow marine

Table 13.1. Mammal rabble

Major Groups	Minor Groups and Characteristics	
Morganucodontans†	Small mammaliaforms; teeth with three simple cusps	
Haramiyidans†	Mammaliaforms; teeth with numerous cusps in approximate rows and tapering heights	
Docodonts†	Mammaliaforms; teeth with complex inner and outer cusps of various heights with many shearing surfaces	
Eutriconodonts†	Mammaliaforms; teeth with usually three cusps in an anterior-posterior line, either all subequal height or with a tall central cusp and flanking shorter cusps	
Australosphenids (probably including monotremes)	Mammals; independently derived tribosphenic molars, teeth lost in adult monotremes	
Multituberculata†	Mammals; bladelike posterior premolars, molars with many cusps in anterior-posteriorly aligned rows	
"Symmetrodontans"†	Zhangheotheriids†	Mammals; teeth with three cusps arranged in a triangle in occlusal view
Dryolestoids†	Dryolestids†	Mammals; seven or eight molars with three main cusps and additional cusp
	Paurodontids†	Mammals; three or four molars similar to dryolestids
Zatherians†	Peramurids†	Mammals; first nearly tribosphenic molars
Theria	Eutheria (including placentals)	Mammals; three truly tribosphenic molars
	Metatheria (including marsupials)	Mammals; four truly tribosphenic molars

Note. Many of the mammal group names encountered in this chapter may be unfamiliar, even if the groups are not extinct. This is a guide to some of the terms used in this chapter. † = extinct groups

conditions. The coals represent the swamps and indicate an incredible diversity of animals, but the stars of this fossil collection are the mammals.

Mesozoic Fur-Ball Origins

The mammals of the Mesozoic appeared in the Late Triassic and include some truly unusual groups, but the Mesozoic was the time that each of the major modern lineages appeared, so that any one dinosaur time in the Cretaceous featured archaic or relatively modern mammalian forms (sometimes a mixture of both) scurrying around the burrows, bushes, treetops, and underfoot. Unusual? Docodonts and dryolestids had up to eight molars in each jaw—that's in addition to incisors, canines, and premolars. Most modern mammals have three or four. Multituberculates became extinct in the Oligocene, well into the Age of Mammals, but they occurred throughout the second half of the Mesozoic and unfortunately left no descendants. Despite their lack of a close relation to any living or extant group of mammals—in the skull and dentition, especially—they were superficially rodent-like mammals that may have filled a similar ecological role before possibly being competitively replaced by rodents later on. Haramiyidans were broadly similar to multituberculates, with a few oddities, and were apparently related to them, but they were older and ranged from the Late Triassic to the end of the Cretaceous. Some morganucodontans still had several tooth replacements per socket during the life of the animal, like reptiles and early mammalian ancestors and unlike later mammals' (and humans') lone baby and adult sets. Meanwhile, other morganucodontans had relatively primitive jaw structures. Some Mesozoic mammals (*Shuotherium* and *Pseudotribos*) had lower cheek teeth with a cusp structure almost perfectly anterior-posteriorly reversed from that of most mammals, suggesting that characteristic molar morphology of mammals may have evolved independently multiple times. Morganucodontans and several other groups had small bones that functioned in both their ears and lower jaws (more on that later). Eutriconodonts and "symmetrodontans" had teeth with often just three simple sharp or blunt cusps in a line or triangle. And the odd Mesozoic mammals were present not just early on in prehistory. The docodonts had tooth cusp arrangements unlike any other Mesozoic (or living) mammal, and most are Middle Jurassic, with a few in the Late Jurassic and one ranging into the Early Cretaceous. As noted above, multituberculates ranged well into the Cenozoic, and dryolestoids and eutriconodonts also ranged in the Cretaceous.

More modern mammal groups, conversely, range farther *back* in time than you might expect. Fossil monotremes (i.e., ancestors of the duck-billed platypus and echidna) go back to the late Early Cretaceous, but their origins were probably in the Middle Jurassic. The line leading to modern placental and marsupial mammals (Theria) would have split from monotremes around the same time, and the two therian groups probably split from each other by the Late Jurassic. By the Late Cretaceous, both were common in mammal faunas, although marsupials sometimes dominate fossil samples, even in places like North America that have few marsupials today.

The mammalian fauna from Leiria's Guimarota mine includes the docodont *Haldanodon*, 11 genera of multituberculates, three types of dryolestids, and the paurodontid dryolestoid *Henkelotherium*, known from a nearly complete skeleton (fig. 13.1A). The other paurodontid from the deposit, *Drescheratherium*, is a "sabertoothed" form with enlarged canines. There is also the basal zatherian mammal *Nanolestes*, which is derived—about as close to modern placentals and marsupials (Theria) as you can get in the Late Jurassic.

At the same time, in North America, the Morrison Formation was being deposited and mammal remains were being buried in places like Como Bluff's Quarry 9, the Fruita Paleo Area, and Dinosaur National Monument, among many others. The mammal fauna of the Morrison is a bit less abundant in fossil numbers but is more diverse in species than Guimarota. There are in the Morrison a docodont, *Docodon* (fig. 13.1*B*); fewer genera of multituberculates, *Ctenacodon*, *Psalodon*, *Glirodon* (fig. 13.1C), *Zofiabaatar*, and *Morrisonodon*; several dryolestids like *Dryolestes*, *Laolestes*, and *Amblotherium* (fig. 13.1*D*); and several types of paurodontids, "symmetrodontans," and eutriconodonts. Although Guimarota and the Morrison do not share any genera of mammals in common, several Morrison forms also occur in the Middle or Late Jurassic of the UK, *Amblotherium* and *Ctenacodon* among them. The dryolestid jaw that hooked me all those years

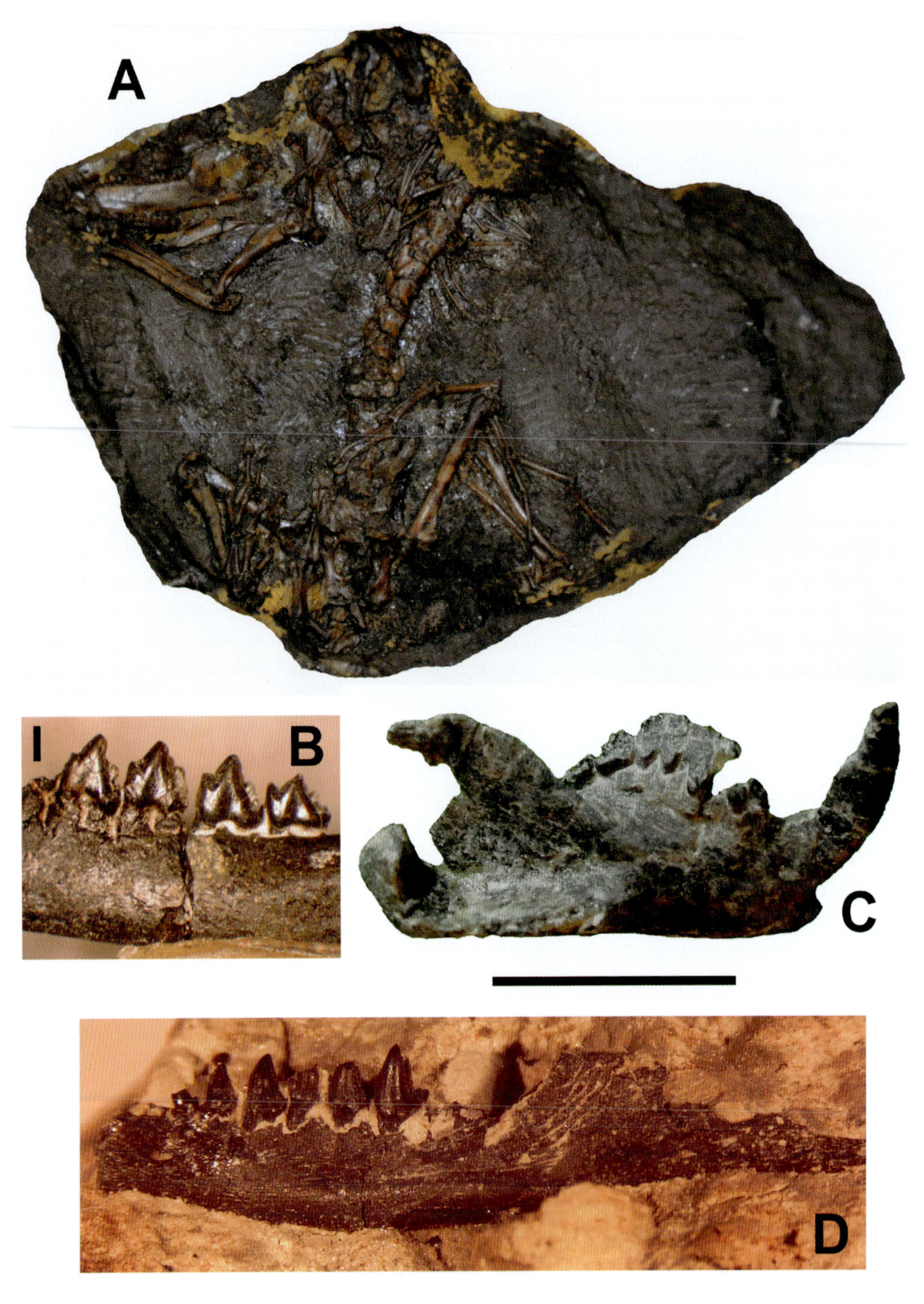

13.1 Some Late Jurassic mammals. (*A*) The Late Jurassic paurodontid *Henkelotherium* from the Guimarota Mine site in Portugal. (*B*) Close-up of four of the molars of the Late Jurassic docodont *Docodon* (SDSM 60480) from Wyoming, US. (*C*) Lower jaw of the Late Jurassic multituberculate *Glirodon* from the Fruita Paleontological Area in Colorado, US (LACM specimen). (*D*) Lower jaw of the dryolestid *Amblotherium*, Morrison Formation, Wyoming, US (SDSM 148545). *Photo in A courtesy of Octavio Mateus; all other photos by author.*

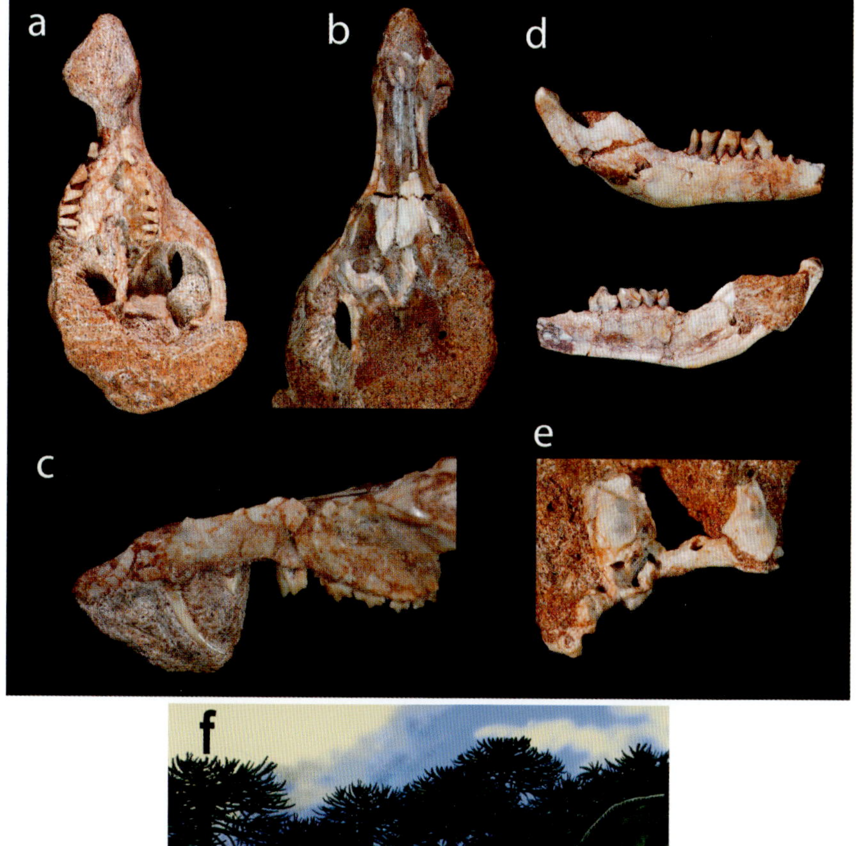

13.2 The dryolestoid mammal *Cronopio* from the Late Cretaceous–age Candeleros Formation of Argentina (MPCA 454). (*A–C*) Skull in ventral (*A*), dorsal (*B*), and left lateral (*C*) views, showing narrow snout and enlarged canine tooth. (*D*) Lower jaw in labial (*top*) and lingual (*bottom*) views. (*E*) Detail of skull base and ear region. (*F*) Reconstruction of the living animal by Jorge Gonzalez. *All images courtesy of Guillermo Rougier.*

ago in Wyoming (see chap. 1) proved to be a relatively large, but probably juvenile, individual of *Amblotherium*.

Among specialized mammalian "sabertooths" from the Mesozoic are the aforementioned *Drescheratherium* from Guimarota; an unstudied, single *Drescheratherium*-like snout cross section from the Morrison Formation at the Fruita Paleo Area; and then the nearly complete skull of *Cronopio* from the Cretaceous of Argentina (fig. 13.2), a small dryolestoid whose snout bears a striking resemblance to that of "Scrat" the acorn-obsessed mammal from the *Ice Age* movies. I wish I could say I was kidding, but it's uncanny. For that matter, *Drescheratherium* looks a bit like "Scrat" too, as once noted in a talk by Universidade Nova de Lisboa paleontologist Octavio Mateus. Even the Early Cretaceous dryolestid *Phascolestes* from England sported some good-sized canines. What these forms were doing with their enlarged canines isn't immediately obvious. Were they showing off to impress or intimidate? Piercing insect exoskeletons? We can't be sure. But the enlarged canines do make the animals rather distinctive.

But that is just the beginning. Mesozoic mammals occur all over the world, and they range from the Late Triassic through the end of the Cretaceous, the end of which they survived, as a group. More than two-thirds of mammalian history occurred alongside dinosaurs in the Mesozoic! Particularly productive over the years have been the Late Cretaceous deposits of western North America and Asia. But the Middle Jurassic to Early Cretaceous lake deposits of China have proven perhaps most informative morphologically.

As fossils, mammals are distinguished primarily by their tooth and jaw structure, as we will see, and the group evolved from cynodont therapsids among the synapsids. Synapsids evolved during the Paleozoic. The sail-backed *Dimetrodon* from packs of "dinosaurs" you get at the grocery store? That's a synapsid, not a dinosaur. Synapsids split off from the reptile line very early. After the first amphibians appeared on land, one of the most important developments was the egg of amniotes, which allowed these first nonamphibian tetrapods to lay eggs in drier habitats away from water. The amniotic egg is characterized, in addition to usually having a hard shell (think breakfast), by the presence of an amniotic sac surrounding the embryo, which is what helps keep the young one from drying out. Amniotes with live birth (like humans) have the amniotic sac but have ditched the shell and the laying of the eggs. So all amniotes today, reptile, bird, and mammal alike, are related by this shared characteristic, among others, and this appeared well before the Mesozoic. The amniotes then split rather quickly into the reptile line, including the ancestors of dinosaurs and birds, and the synapsid line, which led eventually to mammals. During the late Paleozoic, the synapsids were represented by things like the sail-backs *Dimetrodon* and *Edaphosaurus*, the caseasaurs like tiny-headed *Cotylorhynchus*, the dinocephalians such as *Titanophoneus* and *Moschops*, the saber-tooth gorgonopsians like *Smilesaurus*, and dicynodonts like *Lystrosaurus*. These small dicynodonts ranged into the Early Triassic, and by the Late Triassic, dicynodonts had only gotten bigger in forms like *Placerias* from the Chinle Formation in Arizona (see chap. 1).

During the Permian, therapsid synapsids represented the lineage that eventually led to mammals. Among the therapsids, a group known as the cynodonts gave rise to the ancestors of mammals. Cynodont therapsids in general, including the ancestors of mammals, probably had hair and probably laid eggs (like monotreme mammals), although this has not yet been confirmed through any fossil egg finds. By the Middle to Late Triassic, some cynodonts, such as the South American *Brasilitherium*, were small and probably had hair. Also by the Late Triassic, the tritylodontids, the sister group to mammals (or, more precisely, Mammaliaformes; we'll get into that later) lived alongside dinosaurs. Tritylodontids were relatively large and had blunt, rectangular grinding teeth with two or three rows of V-shaped cusps; large, almost tusklike teeth at the front of the mouth were enlarged incisors, and the animals lacked canines. They are hypothesized to have been largely herbivorous generalist feeders that probably fed on a minor component of insects and other invertebrates.

Crews working in the Lower Jurassic Kayenta Formation in Arizona unearthed three tritylodontids in *Kayentatherium*, *Oligokyphus*, and *Dinnebitodon*, among other vertebrates such as the dinosaur *Dilophosaurus*, in the Ward Terrace region. *Kayentatherium* was about the size of a medium dog and was a stocky animal with a relatively long torso but short legs and tail and a robust skull with tusks and grinding teeth. It may well have had hair and appears to have had litters of up to three dozen pups. Tritylodontids are fairly widespread and long ranging in the Mesozoic, for nonmammalian synapsids, with forms from the Early Jurassic of China, Germany, South Africa, and Antarctica (the latter being where they cooccur with the "frozen-crested" theropod *Cryolophosaurus*); the Middle Jurassic of the UK; and the Early Cretaceous of far northeastern Russia.

What I refer to here as mammals are more precisely known as Mammaliaformes, the Mammalia itself being restricted to all the descendants of the common ancestor of monotremes and therians (including modern placentals and marsupials). So groups like morganucodontans, docodonts, and haramiyidans would be mammaliaforms that are outside Mammalia. But some ancient mammal groups, such as eutriconodonts, dryolestoids, and multituberculates, would be within Mammalia. When I refer to mammals in this book, I am speaking of the more inclusive Mammaliaformes.[1] Ecologically and physiologically, these were all stem mammals, and their positions relative to modern taxa and the strict Mammalia are less important. If we took a time machine back to the Triassic or Jurassic and watched a stem mammal such as a morganucodontan go through its furry, daily routine, we'd be unable to recognize much difference between them and taxa included formally within Mammalia. So mammaliaforms will be mammals here.

Of course, when we think of characteristics of mammals today, we think of hair, mammary glands, and warm-bloodedness. We less often think of differentiated teeth with specific dental formulas and replacement patterns, postdentary bones that shrink and migrate into the middle ear, or epipubic bones, the things that can actually fossilize (although, if you're lucky, hair will do so now and then). Fossil mammals can be recognized by teeth that are specialized for specific jobs, from the nipping incisors to the stabbing canines to the grinding or slicing molars and premolars, a tooth differentiation (not entirely unique to mammals) known as heterodonty. And unlike reptiles, most mammals, like humans, only get two sets of teeth, "baby" teeth and a permanent set.[2] The epipubic bones of the pelvis are lost in many more recent mammals but were quite common among early mammals and in marsupials; these bones are also present in some tritylodontids and appear to help provide leverage to the trunk muscles that stabilize the body core.

Jaw Story

The story of the mammalian lower jaw and ear is a long one that shows the modification of old synapsid traits into the specialized ones of mammals. The more that has been found from around this transition, the more the cynodont-to-mammal distinction has become a little blurred. Let's start at the end members and then work our way through what happened to get from one point to the other. In most synapsids (and reptiles), the lower jaw is composed of multiple individual bones and the ear of a bone called the stapes. In mammals like humans, the lower jaw is a single bone, the tooth-bearing dentary, and the middle ear has three bones (stapes, incus, malleus), which transfer energy from the eardrum to the inner ear where the vibrations are sensed. In earlier synapsids and reptiles, there are a number of bones posterior to the tooth-bearing dentary. Most of these were gradually lost in the transition to mammals, and one of those postdentary bones was incorporated into the middle ear: the quadrate of the skull became our incus and the articular of the lower jaw became our malleus. Now, in reptiles and earlier synapsids, the joint between the skull and the lower jaw is between the quadrate (in the skull) and the articular bone (in the lower jaw). As these transitioned into the middle ear in mammals, they could no longer serve as that articulation point, and so in humans and other mammals, the lower jaw-skull articulation is between the dentary and the squamosal bones.

The fascinating thing about the host of early mammals is that you can see this transition occurring in the fossils. As with the embryology and ontogeny of modern mammals (especially marsupials), we can essentially "watch" as the postdentary bones of cynodonts shrink even more, begin to form the jaw joint, and then migrate into the middle ear in fossil mammal taxa. In some early mammal forms and Late Triassic cynodonts like *Diarthrognathus* (which literally means "two-jointed jaw"), the jaw joint is actually formed by both sets of bones: the dentary and articular side by side in the lower jaw, articulating with the quadrate and squamosal in the skull. This is a transitional phase. As the Mesozoic goes on, mammals form a more modern arrangement of the ear, and more modern (crown-group) mammals appear, becoming more like today's mammals in other aspects too.

The earliest mammals (mammaliaforms) were the Late Triassic–age *Adelobasileus* from the Tecovas Formation of west Texas; relatively little of this form is known, however, just the back of the braincase. A similarly primitive but geologically younger genus, *Sinoconodon* from the Early Jurassic of China, appears to have continued to replace its teeth throughout its lifetime (the canines up to four times!), a tendency that correlates with indeterminate growth of the skull. On the other hand, the jaw joint in *Sinoconodon* is a bit more advanced than we might expect for a mammal with more primitive tooth replacement.

The morganucodontans from Wales, China, North America, Africa, and Europe included *Megazostrodon* from the Late Triassic of Lesotho, *Dinnetherium* from the Lower Jurassic Kayenta Formation of Arizona, and *Cifellilestes* from the Upper Jurassic Morrison Formation of Utah, among many others. These mammals were small, probably nocturnal and insectivorous, and had long slender jaws with relatively simple, multicusped teeth. Some morganucodontans have both a dentary-squamosal and articular-quadrate jaw joint. As mentioned above, this process started in derived cynodonts. *Morganucodon* itself represents a medium stage in the jaw-ear transition: the main joint was formed by the dentary and squamosal, while the quadrate and articular separated partially and functioned in hearing. The reduced articular was still slightly connected to the lower jaw. Various stages of these partially developed dual jaw joints are present in other mammals such as docodonts and dryolestoids through the Jurassic.

In *Yanoconodon*, the ear bones are reduced further but are still connected to the lower jaw by an ossified Meckel's Cartilage. By the time of *Origolestes*, a zhangheotheriid from the Early Cretaceous of Liaoning, China, the ear bones are fully detached from the ossified Meckel's Cartilage[3] but are not quite as small as those of modern mammals such as *Monodelphis*, the short-tailed opossum.

This change in jaw and ear structure in mammals was related in part to the more active role that chewing plays in mammals. The complex teeth of mammals indicate many cutting and grinding surfaces and more complicated and thorough chewing than in dinosaurs, for example. Reducing the lower jaw to just the tooth-bearing dentary allowed mammals to move the jaw joint closer to the tooth row and thus have more mechanical advantage. Also, once mammals had three little ear bones instead of just one, they could hear more easily and have access to much-higher-frequency sounds than are available to other tetrapods, which ushers in abilities like echolocation, later seen in bats and whales.

Ecology and Phylogenetic Diversity

The cynodonts of the Late Permian to the Late Triassic, at least those close to mammals, were probably hairy and may well have been nocturnal or crepuscular, the latter meaning active during morning and evening twilight.

Some may have been fossorial (digging), and these traits were probably passed on to the earliest mammals. *Prozostrodon*, a protomammalian cynodont of the Middle–Late Triassic of Brazil, was about the size of a house cat and likely a predator of small reptiles; it may have been nocturnal.

Mammals had diversification events in the Late Triassic to the Early Cretaceous among morganucodontans and other early forms, then in the Middle Jurassic among eutriconodonts, "symmetrodontans," and multituberculates. Another diversification in the Late Jurassic occurs among eutriconodonts, multituberculates, dryolestoids, and peramurids. The metatherians and eutherians (broadly, the marsupials and placentals, along with the fossil relatives of each, respectively) diversified greatly in the Early Cretaceous, followed by the monotremes and relatives (australosphenidans). Finally, the eutherians underwent another radiation in the Late Cretaceous, even though most of the Mesozoic mammals in Late Cretaceous rocks in the western United States tend to be multituberculates and metatherians.

The family tree of mammals is diverse, and among these early ones, we see a range of morphologies and ecologies. Morganucodontans mammals were small and mostly insectivorous; they had long, slender jaws with relatively simple sharp-cusped teeth. The docodonts, on the other hand, had teeth with a complex arrangement of cusps and shearing surfaces, all at different heights and orientations—these are among the oddest mammal teeth out there. Haramiyidans had multicusped teeth with the cusps arranged in rows, combined with a relatively primitive lower jaw. Eutriconodonts have three main cusps per tooth in a line, generally either all of the same height or with the central cusp distinctly taller than the other two, while "symmetrodontans" also have three cusps with a central one taller than the other two, only instead of being in a line, the cusps form a triangle when viewed from above (for the lower teeth).

Multituberculates are rodent-like in having pairs of long, gnawing-type incisors in the front of the jaw, no canines, and a set of premolars and molars built for grinding and shearing. These latter teeth (molars and upper premolars) are generally squarish to rectangular with either two or three rows of low cusps in a line (thus the "multitubercles" of their group name). The unique part of multituberculates is their last premolar (or premolars, in early forms), which is shaped like a half circular-saw blade, complete with striations and ridges, set in the jaw. The jaw itself was of a fairly modern grade, with only the dentary present and a derived middle ear with incus and malleus.

Dryolestoids, including dryolestids and paurodontids, have a triangular three-cusped arrangement of their molars similar to the "symmetrodontans," but they also have cusps coming off the posterior end of the tooth, the very beginnings of what are known as a "talonid basin" and a "tribosphenic" molar seen in modern mammals. This development is further advanced in zatherians such as *Peramus* and *Nanolestes*, and in function is somewhat like a mortar-and-pestle arrangement with the talonid basin as the mortar. The australosphenids, including monotremes, are largely from southern continents and also have tricuspid teeth with the cusps in a triangle, but this arrangement may have developed independently in this group. Although modern monotremes are toothless as adults, young platypuses, along with fossil monotremes and some australosphenids, have these characteristic teeth. In therian mammals, including the marsupial and placental lines, the tribosphenic molars and talonid basins are fully developed.

But what we are really learning from the mammals of the Mesozoic is that, despite their average smaller size, their ecologies were nearly as diverse as what we see today. Some were general terrestrial carnivores and omnivores similar to the opossum or raccoon; these include *Morganucodon*, *Yanoconodon*, and *Zhangheotherium*, among others. Others were specialized for burrowing and eating ground-dwelling insects like termites and ants, with spade-like front claws and robust humerus bones and large flanges on the ulna for leverage of the digging muscles, along with peg-like teeth lacking enamel, forms that are now imitated by animals like moles, aardvarks, and armadillos. The fossil mammal *Fruitafossor* from the Morrison Formation was one of these pioneer burrowing, insect-eating specialists, albeit a small one. Another burrowing specialist was the docodont *Docofossor* from the early Late Jurassic of China. Then there is the semiaquatic, paddle-tailed docodont *Castorocauda*, from the Middle to the Late Jurassic of Asia. Although morphologically beaver-like with its flat tail, *Castorocauda* probably had a far more piscivorous diet than a beaver and may have been more ecologically equivalent to a river otter. Other semiaquatic Mesozoic mammals that may have been ecologically similar to modern forms such as desmans, muskrats, beavers, and river otters include the docodont *Haldanodon* from the Late Jurassic of Portugal and possibly *Docodon* from the Morrison Formation, although both of these docodonts may have been burrowers to some degree also.

Scansorial mammals live on the ground but spend a significant amount of time scampering into, up, and out of bushes and trees; they are neither strictly ground-dwelling nor arboreal but can climb well when they want to. Tree shrews and those acrobatic squirrels, tightroping it on power lines above barking dogs and swirling around, raiding bird feeders in many a backyard, are examples of scansorial mammals. These forms were beat to such agile antics in the trees and bushes by forms like the Late Jurassic paurodontid *Henkelotherium* from Guimarota, the Early Cretaceous eutherian *Eomaia* (fig. 13.3), and metatherian *Sinodelphys* (fig. 13.4B), the latter two from Asia. Many multituberculate mammals were probably scansorially adapted as well. And truly arboreal (tree-dwelling) forms were present in the Mesozoic in the Middle Jurassic haramiyid *Arboroharamiya* and the docodont *Agilodocodon* (fig. 13.5A) (both from China) and possibly in *Henkelotherium*, if it was more strongly adapted for climbing, among a number of others.

Large, gnawing incisors, no canines, and grinding molars are characteristic of rodents today. During the Mesozoic, as we have seen, this was a combination independently arrived at by multituberculates, including

13.3 Ecological specialization in Mesozoic mammals I. Part and counterpart of the scansorial eutherian (stem-placental) mammal *Eomaia* (CAGS 01-IG1-a,b) from the Early Cretaceous of China. *Images courtesy of Zhexi Luo.*

the very rodent-like *Rugosodon* from the early Late Jurassic of (once again) China. Lagerstätten in China have clearly been going crazy with new taxa in the past 30 years and, as we have seen, not just among mammals.

Even the flying squirrel, colugo, and sugar glider are doing nothing original. During the Middle to Late Jurassic in Asia, the haramiyidan *Maiopatagium* (fig. 13.5*B*) and the eutriconodont *Volaticotherium* glided between trees using skin between their limbs and body, just like the modern forms, only 160 million years ago.

Most Mesozoic mammals were small; in the Late Jurassic–age Morrison Formation, *Docodon* was the largest, probably around 140 g (5 oz), possibly less. During the Early Jurassic, the primitive *Sinoconodon* may have gotten up to 500 g (18 oz), and *Castorocauda* may have hit 700 g (25 oz) in the Middle Jurassic (fig. 13.4*A*). *Cifelliodon* from the Lower Cretaceous Cedar Mountain Formation in Utah grew up to an estimated 1.3 kg (2.9 lbs). But by the Cretaceous, some others were getting significantly larger. *Didelphodon*, a Late Cretaceous carnivorous metatherian, weighed up to 5 kg (11 lbs) and may have preyed on taxa as large. The heavyweight champion among Mesozoic mammals, however, is the Early Cretaceous eutriconodont *Repenomamus*, which had a jaw over 10 cm (4 in) long, reached up

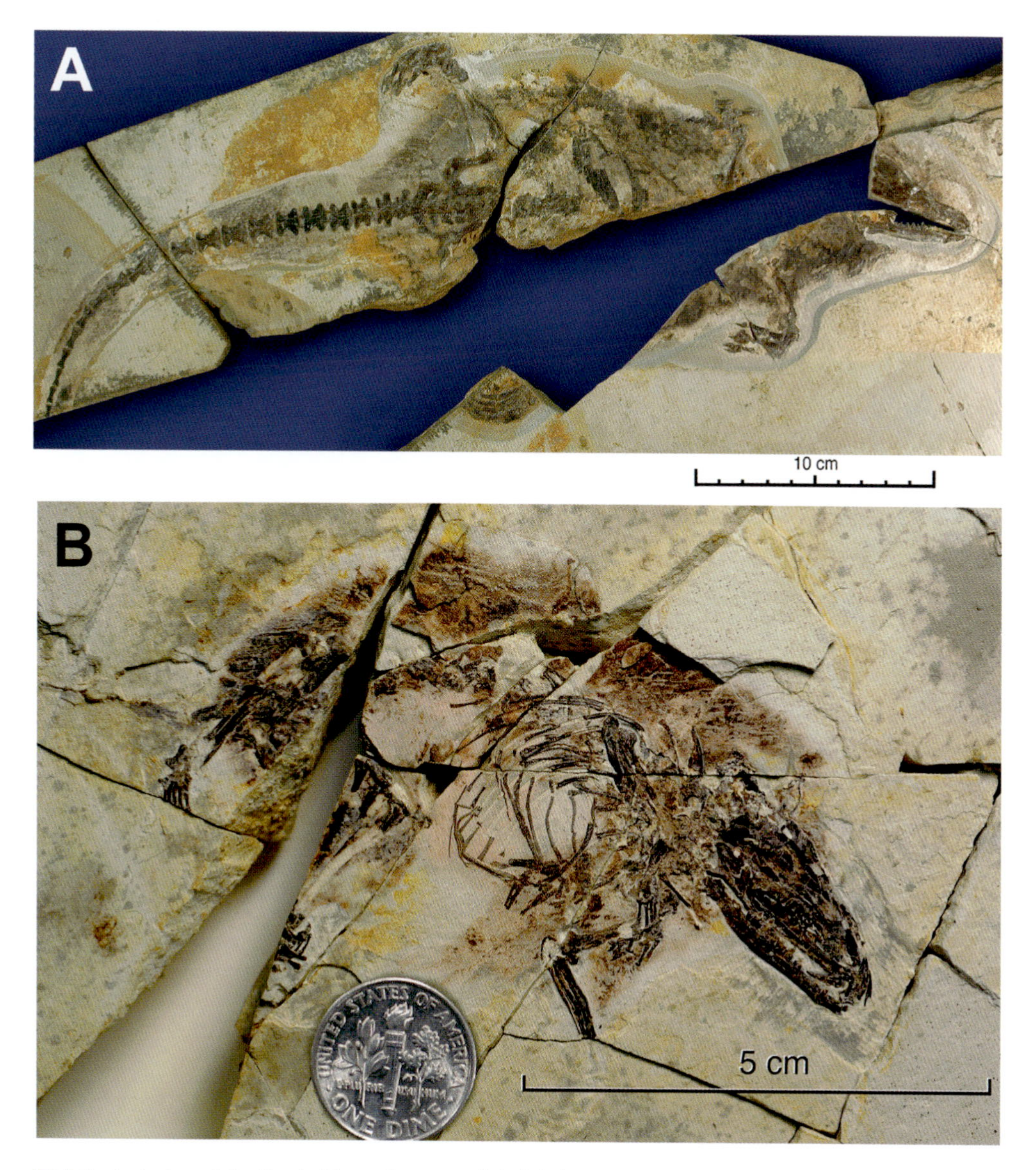

13.4 Ecological specialization in Mesozoic mammals II. (*A*) The semiaquatic, "beaver-tail" docodont *Castorocauda* (JZMP 04117) from the Middle to Late Jurassic of China. (*B*) The scansorial (tree-, bush-, and ground-scampering) metatherian (stem-marsupial) *Sinodelphys* (CAGS 00-IG03) from the Early Cretaceous of China. *Images courtesy of Zhexi Luo.*

to 12 kg (26 lbs), and is known from stomach contents to have fed on baby dinosaurs. No more cowering in the shadows all the time—mammals *ate* dinosaurs (albeit small ones) and probably held their own against the scaly, feathery brutes a little better than we usually give them credit for.

As noted above, most Mesozoic mammals were probably nocturnal to crepuscular; that is, they were active mainly at night or during the morning and evening twilight hours. It was safer then. But we must avoid the temptation to assume that all Mesozoic mammals were only nocturnal or crepuscular and that it was only safe for them at night or, conversely, that it was

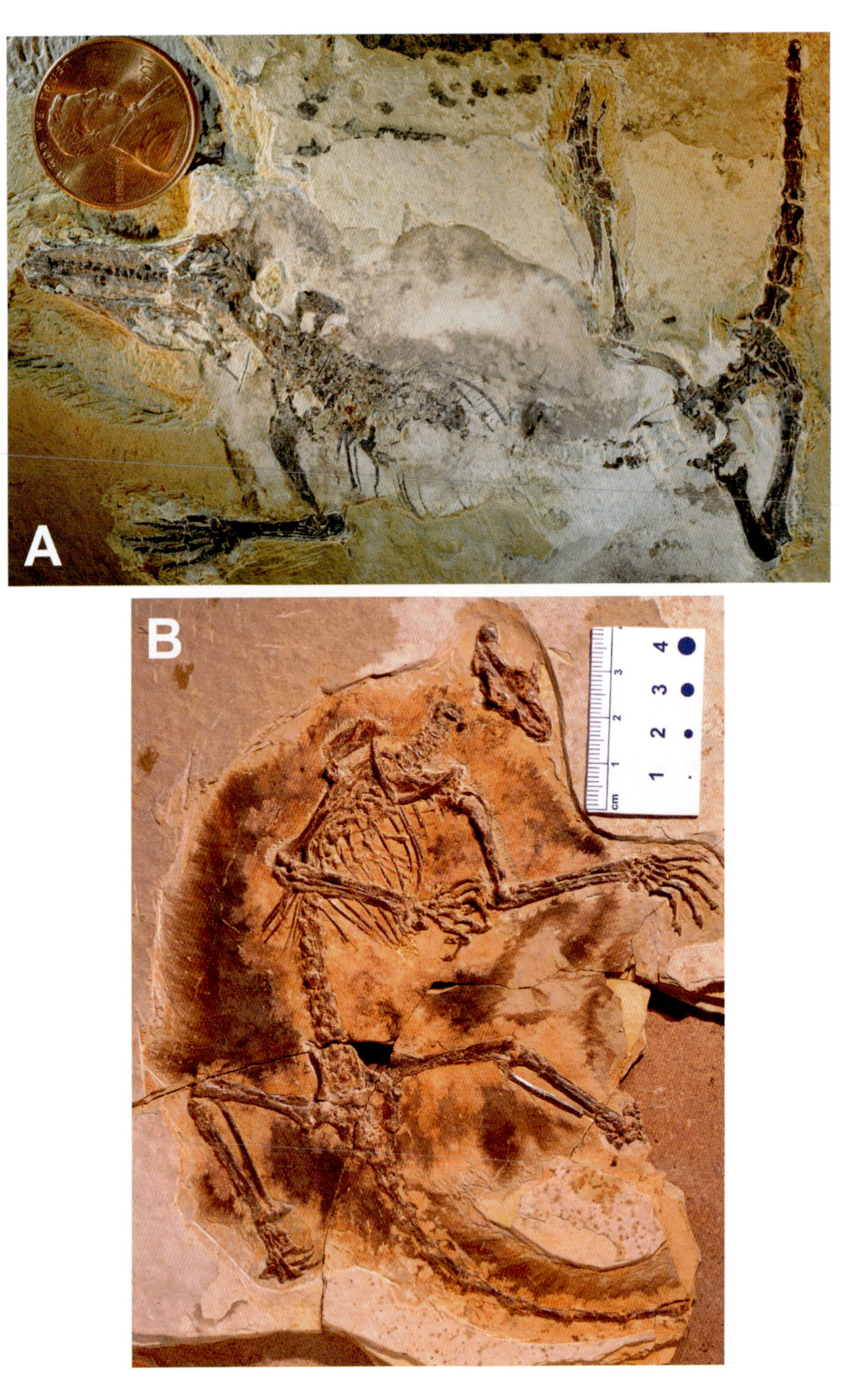

13.5 Ecological specialization in Mesozoic mammals III. (*A*) The Middle Jurassic tree-climbing docodont *Agilodocodon* (BMNH 1138) from China. (*B*) The gliding haramiyid *Maiopatagium* (BMNH 2940) from the Middle to Late Jurassic of China. *Images courtesy of Zhexi Luo.*

entirely safe for them at night. Undoubtedly, a few mammal species would have ventured out during the day and survived, and just as assuredly, some dinosaurs discovered that there was good hunting in the evenings. Some small theropod dinosaurs, such as *Ornitholestes* during the Late Jurassic, had orbits that were unexpectedly large for their skull compared to average small theropods, and these may well have been adapted to hunt small mammals and other vertebrates at night. Many mammal characteristics are attributed to the "nocturnal bottleneck," the early evolution of mammals as night-active species and their probable specialization for it for much of the Mesozoic.[4] But there must have been postbottleneck nonconformists on both sides of the competition—some mammal species might have found that things were less crowded among their small foraging peers during the day, and some small carnivorous dinosaur species discovered that nighttime was a smorgasbord of small mammals to hunt. Some mammals found benefits to balance out the risks they were taking by not falling in line with most of their peers, and some dinosaurs found the nighttime dining worth the investment of modifying their species for such nocturnal specialization.

Even the "Island Rule" was apparently loosely functioning during the Mesozoic—not that there is any reason that should be surprising. This is the tendency (more than an actual rule) of small species to become large and large ones to become small when isolated on smaller landmasses. This is represented by island dwarves and giants such as the pygmy mammoths and the giant lemurs of recent millennia. *Adalatherium*, the "crazy beast," was a large (3.1 kg, 7 lb) gondwanatherian mammal (part of the sister group to multituberculates) from the Late Cretaceous of Madagascar. Most multituberculates were rather small, but this relative was pushing *Repenomamus* sizes, probably as a result of the geologically recent (during the Cretaceous) isolation of Madagascar.

By the Late Jurassic, mammals collectively (especially docodonts, eutriconodonts, and multituberculates) had diversified into every major modern mammal locomotion mode except hopping and powered flight. And even hopping was probably present in some Mesozoic mammal species, as suggested by the admittedly equivocal evidence of the trackway ichnogenera *Koreasaltipes* from the Early Cretaceous of Asia and possibly *Ameghinichnus* from the Middle Jurassic of Argentina.

Sometime in the middle of August that year long ago, we packed up our camp outside Rangely, Colorado, loaded all our fossils and concentrate from the Williams Fork Formation into the vehicles, and began the journey back to San Diego. Weeks earlier, when we were inspecting an anthill and trying to decide whether to sample it for screenwashing, Dave Archibald held up a tiny Late Cretaceous mammal tooth with a pair of tweezers and said, "Here you go. Get a good look. It's likely the only one we'll actually *see* all summer."

And indeed it was. Even more so than with the fish, most of the mammals we found that summer only revealed themselves after the screenwashing concentrate was inspected under a microscope back in the lab. To see a Mesozoic mammal in the field there was and is a rare thing, but a decent number turned up there eventually and at other sites since.[5] It just takes work. As it is with dinosaurs, it is with Mesozoic mammals—the number one tool you need is patience.[6]

Copycats: Pervasive Convergence

Mesozoic mammals seem to have been doing almost everything in the small niches that modern small mammals do. They did not become the truly large browsers, grazers, and top predators of the time that they did during the Age of Mammals—that role was filled by the dinosaurs during the Mesozoic. But this ecological diversity among Mesozoic mammals, really only revealed in all its impressive glory in the past 30 years or so, means that in some sense, our modern, familiar animals are largely copycats. We shouldn't be surprised that Mesozoic mammals were flying, jumping, swimming, or burrowing back then but rather that modern small mammals are still doing many of the same things that mammals have been doing for 200 million years.

There are only so many ways of making a living, and early mammals seem to have come across many of those ways rather quickly, from multituberculates more or less anticipating what we think of as the rodent role to docodonts already evolving arboreality, fossorial (burrowing) habits, and semiaquatic tendencies long before any modern mammals had done these things. This phenomenon of unrelated forms developing similar morphologies as a result of independently occupying similar ecological roles is known as convergence.

Convergence runs all through evolution and operates at many levels, even the molecular. There are so many examples, they fill books. We've already seen it in the similarity of ichthyosaur reptiles with modern dolphins, for example, both being fast-swimming, open-ocean piscivores—they came from totally different branches on the vertebrate tree but ended up looking very much the same because they needed the same things to be most efficient (in this case, high-speed swimming capability for the body and tail and a snout made for catching fish). We've also seen it in the batlike anurognathid pterosaurs and some of the armored and armadillo-like herbivorous and omnivorous notosuchian crocodylomorphs, as well as between various Late Triassic archosaurs and dinosaurs of the Jurassic and Cretaceous (e.g., armored aetosaurs versus armored ankylosaurs).

Similarly, many of the seemingly industrious and creative things small mammals do today—gnawing, gliding, swimming, scampering, hopping, living in trees, eating things as big as you—were already being done by their extinct and often indirect ancestors 160 million years ago or more. The Mesozoic already had the diggers, gliders, large carnivores, tree dwellers, termite eaters, and swimming docodonts with beaver tails, among others—specialists galore. So next time we see a rodent bounding from lawn to fence to tree branch, hoarding a stash of an omnivore's mix of dietary items, or the next time we see a flying squirrel in action in the forest, we can think of the multituberculates and of *Maiopatagium*, respectively, and we can remember of Mesozoic mammals: *They pulled those stunts already, deep in time, and with dinosaur witnesses.*

Notes

1. This distinction has also been made by referring to the more restricted "crown Mammalia" and the more inclusive "total Mammalia," the latter more or less equivalent to Mammaliaformes.
2. Except in the case of adult molars, for which you only get one set that emerges behind the premolars (which themselves replace what had been functional "baby" molars).
3. This is a medially and posteriorly positioned cartilaginous structure around which the bones of the lower jaw develop; its position is indicated in many mammals by a groove along the lingual surface of the dentary.
4. Among the nocturnal-specialization characteristics of early mammals are the reduction of the optic lobes to tiny superior colliculi and the expansion of areas of the cerebrum to handle the integration of smell, hearing, and vision. The color blindness of so many mammals speaks to this as well.
5. Our crews just recently found another multituberculate specimen in the Williams Fork Formation that we hadn't noticed initially in the field.
6. Patience is the main tool needed for all of vertebrate paleontology, really. You can't rush it in the field or the lab. This can be quite true in some cases with invertebrate paleontology as well.

Epilogue

A Bump in the Road and Beyond

14

OUR CAST OF CHARACTERS changed a bit over the millennia, species came and went, but otherwise, as groups, our new friends were sailing along until the last day of the Cretaceous. The story of the end of the Cretaceous by bolide impact has been told enough times; it doesn't need another telling, least of all by yours truly, as I am more comfortable about 85 million years earlier in the Jurassic. Enough sensational bad-day stories have been written about the Cretaceous–Paleogene (K–Pg) boundary event that I couldn't stomach adding to the collection even if I were tempted to cover it here, other than to reiterate that the heroic nonavian dinosaurs around the world eventually met their unfortunate demise as a result of the environmental disruption of this event, except locally in North America and presumably some of South America, close to the crater. In these regions, many populations likely were killed within hours as a direct result of the impact's effects. Nevertheless, the beast companions, at least most of them, soon found themselves more or less solo among terrestrial-freshwater vertebrates. Among the major groups we have covered here that were around in the last days of the Cretaceous, only the nonavian dinosaurs, pterosaurs, plesiosaurs, mosasaurs, and ammonoids were entirely missing by the first years of the Cenozoic.

But what happened on land and in the ponds to the groups we've covered here? Although populations were devastated, numerous species survived, and most of the broader beast companion groups made it through to today. It was mostly *T. rex*, *Triceratops*, *Edmontosaurus* (fig. 14.1), and other dinosaurs that took it permanently on the chin and never recovered. Everything larger than about a medium-sized crocodile became extinct, and to have been a terrestrial animal rather than aquatic or semiaquatic was bad as well. Insects and other invertebrates, fish, amphibians, and turtles all lost species but came through in decent shape. Sphenodontids lost some diversity, seemingly to lizards, during the Cretaceous, so there are fewer by the end of the Cretaceous than there were early in the period. But they obviously survived, or we wouldn't be lucky enough to have the cute tuataras around today. Lizards and snakes also lost diversity but survived. Ichthyosaurs were not around to witness the disaster. Plesiosaurs and mosasaurs became extinct. Choristoderes lived on in *Champsosaurus* in the Paleocene. Crocodiles were reduced a little but survived. Pterosaurs were gone. Among dinosaurs, birds were hit hard but survived. Mammals lost a number of species, more in some subgroups than others, but lived on and really diversified to fill the large niches during the Paleocene.

Stories of Hell Creek

In the Hell Creek Formation outcropping in Montana and the Dakotas, the terrestrial K–Pg boundary is preserved at the unit's contact with the overlying Fort Union Formation, and we can see what happened at that moment in time. This is one of surprisingly few places on the globe where the boundary is preserved for an ancient terrestrial setting—most sections around the world are in marine rocks, so we have a better idea of what happened to ammonoids globally after this disaster. (Spoiler: they all disappeared rather quickly.) All our favorite Hell Creek nonavian dinosaurs disappeared at the boundary: *Tyrannosaurus*, *Triceratops*, *Edmontosaurus* (fig. 14.1), and *Pachycephalosaurus*. Most beast companion groups did not.

Over the K–Pg boundary interval in the upper Hell Creek Formation into the Fort Union, freshwater sharks, skates, and rays were hit hard but

14.1 One of the beasts: the hadrosaurid (duck-billed) dinosaur *Edmontosaurus* from the Hell Creek Formation. These herbivores pushed 10 m (33 ft) in length. *Photo by Kevin Eilbeck, South Dakota School of Mines.*

***Facing,* 14.2** Extinction and survivorship at the K–Pg boundary in the Hell Creek Formation to Fort Union Formation of North America. Percentages indicate proportion of Cretaceous species within each group that became extinct by the earliest Paleocene. Fully aquatic freshwater to slightly brackish water groups in dark blue, semiaquatic freshwater groups in light blue, terrestrial groups in green, flying (to possibly aquatic) birds in yellow-green. Thick vertical black line represents boundary, with Hell Creek taxa to the *left* (Cretaceous) and same species also found in the base of the Fort Union Formation on the *right* (Paleocene). Lateral width of bars is schematic and does not correlate with a representation of scaled time; vertical scale of group bar widths does approximate number of species before and after boundary (see scale in *lower left*). Groups that become fully extinct labeled in red. Note higher extinction rates, even among surviving groups, among terrestrial and flying species than among aquatic and semiaquatic species. Although cartilaginous fish occur in the Fort Union Formation and the group survives the boundary, the species in this Paleocene unit consist of Hell Creek taxa that are likely reworked, plus new species not found in the lower formation. Thus, despite their survival as a group, 100% of Hell Creek species of cartilaginous fish seem to disappear. *(Data updated from Archibald, 1996; Hartman et al., 2002; Wilson et al., 2014; and others. Silhouettes from PhyloPic.org.)*

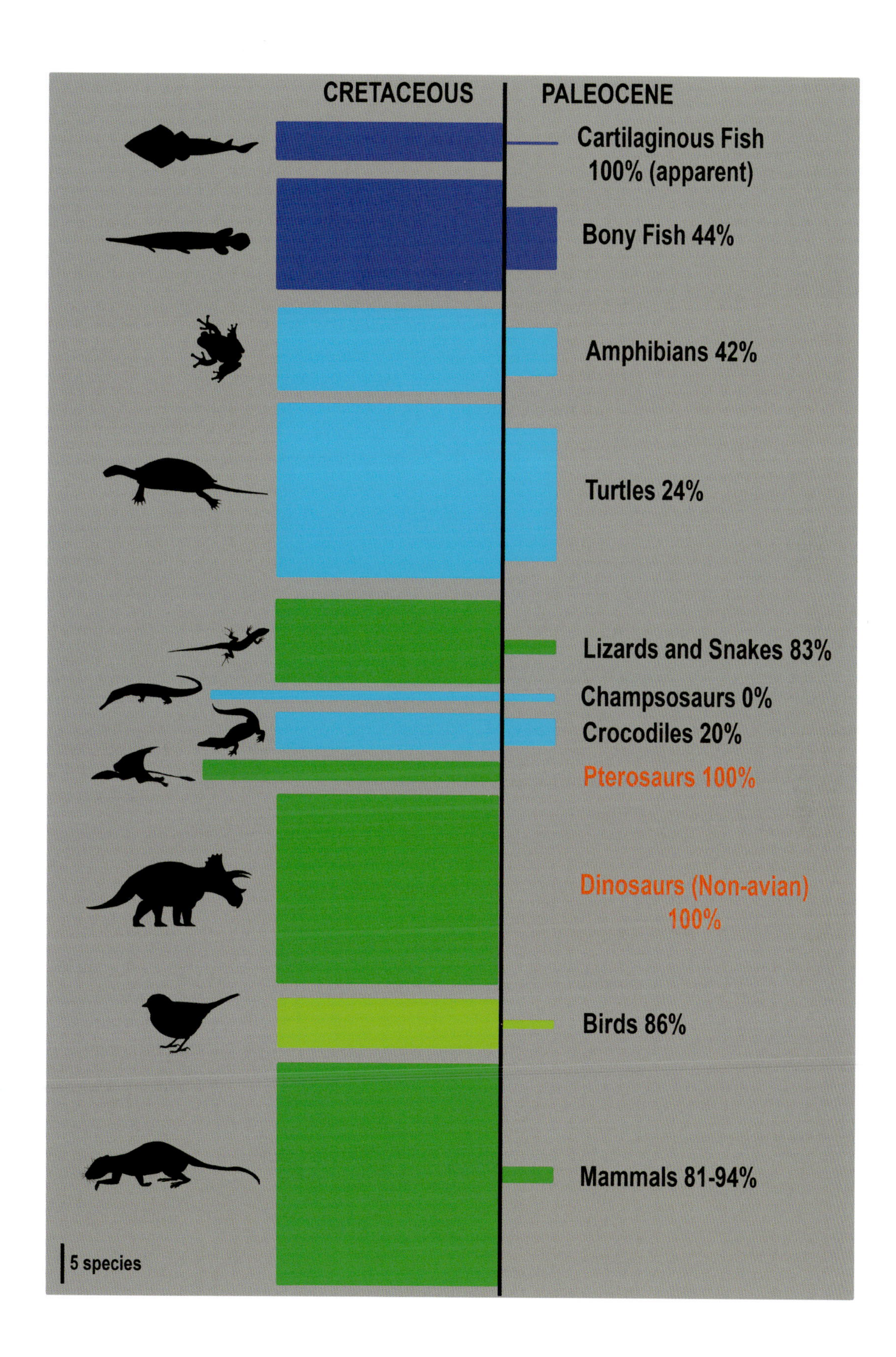
CRETACEOUS
PALEOCENE
Cartilaginous Fish
100% (apparent)
Bony Fish 44%
Amphibians 42%
Turtles 24%
Lizards and Snakes 83%
Champsosaurs 0%
Crocodiles 20%
Pterosaurs 100%
Dinosaurs (Non-avian)
100%
Birds 86%
Mammals 81-94%
5 species

survived; bony fish lost about 44% of species, but 56% survived(!); most frog species survived, but only about 36% of salamanders did (although the majority of lost salamander species disappeared a bit before the boundary itself); and somewhere between 73% and 86% of turtle species *survived* (fig. 14.2). Among lizards and snakes, there is an extinction rate of about 83%. Birds were also hit hard. Nonavian dinosaurs and the pterosaurs disappeared entirely. Meanwhile, in the water, things were not so bad. Most crocodiles survived, and, as we have seen, fish, turtles, and amphibians were not hit as hard (fig. 14.2).

It took about ten million years, but the diversity eventually did recover. The lone choristodere and one of the bigger survivors, *Champsosaurus*, continued into the Paleocene; among crocodiles, 80% of species survivorship is observed; and for mammals, all main groups survived, but they collectively suffered a species extinction rate possibly as high as 94% (fig. 14.2). Still, eutherians, metatherians, and multituberculates all recovered to some degree. Placentals (eutherians) diversified greatly in the following millions of years. Multituberculates hung on until the Oligocene. The case of marsupials (metatherians) is interesting. They were more abundant than placentals in North America during the Cretaceous, but they never really recovered their loss at the K–Pg boundary on that continent, although they certainly did much better in southern continents, especially in Australia where they developed numerous convergent equivalents to placental taxa elsewhere. Monotremes survived, although most of the primitive forms in this group are found earlier in the Cretaceous. Among latest Cretaceous birds in western North America, enantiornitheans, hesperornithines, and ichthyornithines all disappeared for good at the boundary, along with most of the ornithuraeans, but the record of birds in the Late Cretaceous and Early Paleocene is spotty, so it is hard to be sure of these patterns with so few data. What we do know is that some birds (especially ground-dwelling ones) survived across the K–Pg boundary in the Ornithurae and Neornithes and diversified in the Cenozoic. We may not know exactly which species made it, but we know some did.

Vertebrate survival at the K–Pg boundary is associated with two primary factors: (1) being small (less than about 25 lbs, or 11 kg) and (2) being aquatic or semiaquatic. It also appears to have been favorable to have been ectothermic (cold blooded) and a nonamniote (basically, a fish or amphibian). Large, land-dwelling, warm-blooded tetrapods fared particularly badly.

In addition to these animals, plants in the upper Hell Creek Formation survived but were reduced quite a bit in species diversity. There was a 65% extinction rate at the boundary among the plant species, and the postextinction flora was dominated by ferns for some time before other taxa began to come back again. Ferns today often can be among the first plants to reestablish themselves after natural disasters. After this postextinction fern spike, the diversity of plants eventually rebounded during the Paleocene.

After the impact effects of the first few days wore off, after the heat, fire, tsunami, earthquakes, and hailstorms of microspherules that were especially bad in North (and possibly South) America, the amount of dust and debris that clouded the atmosphere probably cut down on sunlight enough to interfere with photosynthesis and cause a global loss of plant species. This in turn hit large herbivorous species even on the other side of the world, which then affected the health of all dependent predatory species further up the tiers of the food web. Within months or years, whole ecosystems would have been severely disrupted around the globe and surviving dinosaur populations would have been doomed.

Meanwhile, the beast companions found themselves with fewer and fewer beasts around until eventually, they found themselves alone. Apparently, small and aquatic species had an easier time insulating themselves from the immediate effects of the disaster and managed to find enough food and shelter in the ensuing ecosystem disruption to come out the other side.

The Rest of the Story

Where the survivors went from those first bleak days of the Paleocene is perhaps a bit cheerier of a story. Freshwater fish of the Cretaceous, the sturgeons, bowfins, lungfish, and gars so common in microvertebrate site fossil samples, chugged right along and are with us today (fig. 14.3*A* and 14.3*B*). Frogs and salamanders sailed through to their current marvelous diversity of color, sound, and resilience, but albanerpetontids got all the way to the Plio–Pleistocene before their extinction. Turtles are everywhere throughout the Cenozoic and today (fig. 14.3*C*), and their high survivorship of the K–Pg boundary may not be surprising, given that kind of tenaciousness. Lizards have obviously done well (fig. 14.3*D*), despite their apparent setback at the end of the Cretaceous. Their record isn't great in Cenozoic rocks, but that is likely a matter of delicate, small bones being hard to preserve—their diversity was likely greater than what we are seeing. And today, their diversity of form and ecology is profound. Champsosaurs, represented by Cenozoic forms such as *Lazarussuchus* (fig. 14.4), hung in there until the Miocene about 15–20 million years ago (fig. 1.3).

Crocodiles (speaking broadly of the modern order, which includes the alligator, crocodile, and gavial families) are widely distributed around the belt of the tropics today, from the alligators and American crocodiles of the golf courses of Florida and the river mouths of Jamaica (fig. 14.3*E*) and Cuba to South America, Africa, Asia, and Australia. And through the Oligocene, they were even more pervasive, being found as fossils in the Paleocene of Wyoming and the Arctic and even South Dakota as recently as 30 million years ago.

As the world began to cool starting in the Oligocene, the latitudinal range of the crocodiles began to compress slowly toward the equator. The tropics were shrinking. But it wasn't until the ice ages, starting 2.5 million years ago, that crocodiles' range really became restricted to what we see today. On that thin band around the globe on either side of the equator and up to about 30°N or 30°S, depending on the climate—which in turn depends on the ocean currents nearby—at 33°N in South Carolina, you'll get alligators, but not at 31°N in Baja California. (That's due in part to the warm Gulf Stream coming up from the tropics in the Atlantic versus the

14.3 Just a few of today's descendants of the Beast Companion survivors. (*A*) A Mississippi River–born shortnose gar (*Lepisosteus platostomus*) in its home tank of 20+ years at the Museums of Western Colorado. (*B*) Lungfish. (*C*) Freshwater turtles sunning themselves at the Honolulu Zoo. (*D*) Iguanian lizards of the genus *Pogona*. (*E*) A scene as old as the Mesozoic: an American crocodile (*Crocodylus acutus*) lurks in the Black River of Jamaica. (*F*) An avian dinosaur of the finch variety. (*G*) A carnivoran mammal (*Felis catus*). *All photos by author.*

14.4 The survivor champsosaur *Lazarussuchus* (BDL 1819) from the Cenozoic (Age of Mammals) of Europe. *Image courtesy of Susan Evans.*

cold California Current coming down from the Alaska along the West Coast—and it's a bit drier in Baja.) During peak ice intervals of recent ice ages, the latitudinal range of crocodiles likely was compressed a bit more.

Birds, the avian dinosaurs, survived. And today, they number close to 10,000 species. Ten thousand colorful, ecologically diverse, feathered dinosaurs of a range of sizes from grams up to hundreds of pounds. Finches (fig. 14.3*F*), ostriches, hummingbirds, penguins, woodpeckers, eagles, herons, and sparrows—they all come to us as descendants of the avian survivors of the K–Pg boundary.

Other than the loss of multituberculates during the Oligocene, mammals arguably fared best of all. Monotremes survived through to today's platypus and echidna, and marsupials took over some southern continents. In northern continents, placental mammals took advantage of the sudden lack of nonavian dinosaurs and diversified into thousands of large herbivorous and carnivorous species, all of which eventually led to today's world dominated by big old mammals. Now-extinct groups that appeared 10–20 million years after the K–Pg boundary, creatures like condylarths, creodonts, uintatheres, and brontotheres, all eventually led (though not necessarily directly) to cats (fig. 14.3G), dogs, bears, bison, horses, camels, cows, and elephants. And ancestral forms of many of those latter groups were present in ecosystems by the Oligocene 35 million years ago. By then, the world really belonged to the latter, more conventional groups.

Legacy of the Beast Companions

It is important to remember that, whatever happened at the end of the Cretaceous, it only happened to a relatively small percentage of dinosaur species that had ever existed. By far, the majority of the dinosaurs had been extinct for millions of years already by the time *T. rex*, *Triceratops*, *Edmontosaurus*, and friends had the misfortune of being near the impact site. *Coelophysis* and the relatively few Triassic dinosaurs? Long gone. The sauropod glory days of the Late Jurassic? All species gone for about 85 million years. Even most forms from the dinosaurs' speciation peak in the Early to middle Cretaceous never came anywhere close to seeing the end-Cretaceous disaster.

So no *Brachiosaurus*, no *Parasaurolophus*, no *Velociraptor*—all memories and fossils by the latest Cretaceous. In two senses, then, it wasn't the end of "the dinosaurs." One, dinosaurs survived on in the birds, and two, it was only the end of the species that happened to be living 66 million years ago—thousands of others were extinct by natural means long before that. Keep in mind that throughout Earth's history, most terrestrial vertebrate species last only about one to two million years to begin with, so we can't expect them all to have been around in the last days of the Maastrichtian. Perhaps that is a positive thought? As bad as the K–Pg boundary event was and as extinct as those earlier forms still were, at least they didn't get gunned down by an asteroid. That perspective cheers me up sometimes. At least a little.

Dinosaurs had 160 million mostly good years, then a couple of bad ones, and that was that, sadly. Completely by accident, mammals suddenly found themselves inheriting the Earth. And they didn't squander the opportunity when it presented itself. It became a mammal world. But plenty of other groups also did well, as we have seen. For example, birds diversified quite a bit postapocalypse too. And whatever hit they took at the boundary, fish, amphibians, turtles, lizards, snakes, and crocodiles all recovered well. By 55 million years ago, ecosystems had recovered and the beast companions were flourishing. The ensuing millions of years brought continued success for most of those groups and have led us to the ecosystems we have inherited and of which we now have stewardship.

The Permian–Triassic, Triassic–Jurassic, and Cretaceous–Paleogene boundary extinctions are showing us what can happen with massive inputs of material (dust, gas, both) to the atmosphere. Whether it is volcanic gases, methane release, and simultaneous burning of thick coal deposits by eruptions of unprecedented scale, or huge amounts of dust thrown up in what is basically a supermassive explosion, or associated ocean anoxia or acidity, the domino effect on ecosystems caused by large changes to the atmosphere and oceans can be catastrophic, especially if it happens quickly, as we see at those boundaries.

Many of the groups we have met here have member species that are endangered today for a variety of reasons, some of which may ultimately relate to changes in our atmosphere too (anthropogenic and otherwise), but many of which, regardless, relate to the other activities of our civilization. As great-grandchildren of Cretaceous mammals, we owe it to our fellow beast companions to maintain our current ecosystems for the survival of all the extended alumni of the Mesozoic, including the bird-dinosaurs and others. Most of us as groups appeared around the same time during the Triassic. We're the Class of Two Hundred Million Years Ago. Deep-time perspective on our current situation, the revelation that such ecological challenges have faced Earth's biota before (with a variety of causes), and that "our" collective vertebrate ancestors were resilient, pioneering, and evolutionarily creative—all of this should give us hope that it is not too late. Perhaps we can pull off another 200 million years. Technologically, modern ecosystems are probably still salvageable, and we mainly need the motivation and the societal will—the determination not to unwittingly cause things to become as bad as they have been, on a few occasions and due to unfortunate alignments of factors, in the distant past. The beast companions are depending on us to preserve what we have all inherited—which also means that *we* are depending on us! We would do well to listen carefully to this call from the wild.

Selected Bibliography

Sources for this work consist of a mix of many shorter, narrow-topic scientific papers, too numerous to list here, other primary references and reviews, and my own work and data gathered over the years in the field and lab. As a guide to sources of additional information, I list here only the books and review papers that cover topics most broadly. But I encourage readers to dive into these references and the shorter papers that focus on details of individual topics, usually cited in the reference sections of the reviews listed here. Many of these sources of original data are available online and/or are open access (other than the books, generally). Primary references can be intimidating when they relate to initially unfamiliar topics, but the data is coming straight from the source, the interpretations direct from the researchers themselves, and eventually, things will become more familiar than you might think. Give it a shot!

Lagerstätten, Diverse Deposits, and Group Overviews

These references cover fossil deposits or broad groups so rich that they apply to numerous chapters throughout the book.

Barthel, K. W., N. H. M. Swinburne, and S. Conway Morris. 1990. *Solnhofen: A Study in Mesozoic Palaeontology*. Cambridge: Cambridge University Press.

Batten, D. J. 2011. "English Wealden Fossils." *The Paleontological Association Field Guide to Fossils* 14:1–780.

Chen, P.-J., Y. Wang, and Y.-Q. Wang. 2011. *The Jehol Fossils: The Emergence of Feathered Dinosaurs, Beaked Birds and Flowering Plants*. Cambridge, MA: Academic Press.

Currie, P. J., and E. B. Koppelhus. 2005. *Dinosaur Provincial Park: A Spectacular Ancient Ecosystem Revealed*. Bloomington: Indiana University Press.

Foster, J. R. 2020. *Jurassic West: The Dinosaurs of the Morrison Formation and Their World*. 2nd ed. Bloomington: Indiana University Press.

Ibrahim, N., P. C. Sereno, D. J. Varricchio, D. M. Martill, D. B. Dutheil, D. M. Unwin, L. Baidder, H. C. E. Larsson, S. Zouhri, and A. Kaoukaya. 2020. "Geology and Paleontology of the Upper Cretaceous Kem Kem Group of Eastern Morocco." *ZooKeys* 928:1–216.

Martin, T., and B. Krebs. 2000. *Guimarota: A Jurassic Ecosystem*. Munich: Dr. Friedrich Pfeil.

Martinetto, E., E. Tschopp, and R. A. Gastaldo. 2020. *Nature through Time: Virtual Field Trips through the Nature of the Past*. Cham: Springer.

Panciroli, E., R. B. J. Benson, S. Walsh, R. J. Butler, T. A. Castro, M. E. H. Jones, and S. E. Evans. 2020. "Diverse Vertebrate Assemblage of the Kilmaluag Formation (Bathonian, Middle Jurassic) of Skye, Scotland." *Earth and Environmental Science Transactions of the Royal Society of Edinburgh* 2020:1–22.

Poyato-Ariza, F., and A. D. Buscalioni. 2016. *Las Hoyas: A Cretaceous Wetland*. Munich: Dr. Friedrich Pfeil.

Steyer, S. 2012. *Earth before the Dinosaurs*. Bloomington: Indiana University Press.

Titus, A. L., and M. A. Loewen. 2013. *At the Top of the Grand Staircase: The Late Cretaceous of Southern Utah*. Bloomington: Indiana University Press.

Zhou, Z.-H., and Y. Wang. 2010. "Vertebrate Diversity of the Jehol Biota as Compared with Other Lagerstätten." *Science China Earth Sciences* 53:1894–907.

Chapter 1: Shadows in the Rain

Brusatte, S. 2018. *The Rise and Fall of the Dinosaurs: A New History of Their Lost World*. New York: William Morrow-Harper Collins.

Erwin, D. H. 2006. *Extinction: How Life on Earth Nearly Ended 250 Million Years Ago*. Princeton, NJ: Princeton University Press.

Fraser, N., and D. Henderson. 2006. *Dawn of the Dinosaurs: Life in the Triassic*. Bloomington: Indiana University Press.

Heckert, A. B. 2004. "Late Triassic Microvertebrates from the Lower Chinle Group (Otischalkian–Adamanian: Carnian), Southwestern U.S.A." *New Mexico Museum of Natural History and Science Bulletin* 27:1–170.

Lucas, S. G., and J. A. Spielmann. 2007. "The Global Triassic." *New Mexico Museum of Natural History and Science Bulletin* 41:415.

Padian, K. 1986. *The Beginning of the Age of Dinosaurs: Faunal Change across the Triassic-Jurassic Boundary*. New York: Cambridge University Press.

Stocker, M. R., S. J. Nesbitt, K. E. Criswell, W. G. Parker, L. M. Witmer, T. B. Rowe, R. Ridgely, and M. A. Brown. 2016. "A Dome-Headed Stem Archosaur Exemplifies Convergence among Dinosaurs and Their Distant Relatives." *Current Biology* 26:2674–80.

Sues, H.-D., and N. C. Fraser. 2010. *Triassic Life on Land: The Great Transition*. New York: Columbia University Press.

Tanner, L. H., J. A. Spielmann, and S. G. Lucas. 2013. "The Triassic System: New Developments in Stratigraphy and Paleontology." *New Mexico Museum of Natural History and Science Bulletin* 61:612.

Chapter 2: A Critical Mass

Boardman, R. S., A. H. Cheetham, and A. J. Rowell. 1987. *Fossil Invertebrates*. Palo Alto, CA: Blackwell Scientific Publications.

Charbonnier, S., D. Audo, B. Caze, and V. Biot. 2014. "The La Voulte-sur-Rhône Lagerstätte (Middle Jurassic, France)." *Comptes Rendus Palevol* 13:369–81.

Condamine, F. L., M. E. Clapham, and G. J. Kergoat. 2016. "Global Patterns of Insect Diversification: Towards a Reconciliation of Fossil and Molecular Evidence?" *Scientific Reports* 6:19208.

Grimaldi, D., and M. S. Engel. 2005. *Evolution of the Insects*. Cambridge: Cambridge University Press.

Labandeira, C. C. 2005. "The Fossil Record of Insect Extinction: New Approaches and Future Directions." *American Entomologist* 51:14–29.

Moore, R. C. 1953. *Treatise on Invertebrate Paleontology Part G Bryozoa*. Boulder, CO: Geological Society of America and University of Kansas Press.

———. 1955. *Treatise on Invertebrate Paleontology Part E Archaeocyatha and Porifera*. Boulder, CO: Geological Society of America and University of Kansas Press.

Sanders, D., and R. C. Baron-Szabo. 2005. "Scleractinian Assemblages under Sediment Input: Their Characteristics and Relation to the Nutrient Input Concept." *Palaeogeography, Palaeoclimatology, Palaeoecology* 216:139–81.

Chapter 3: Sweet Delta Dawn

Grande, L. 2010. "An Empirical Synthetic Pattern Study of Gars (Lepisosteiformes) and Closely Related Species, Based Mostly on Skeletal Anatomy: The Resurrection of Holostei." *American Society of Ichthyologists and Herpetologists, Special Publication* 6, *Copeia* 10 (2A): 1–871.

Grande, L., and W. E. Bemis. 1998. "A Comprehensive Phylogenetic Study of Amiid Fishes (Amiidae) Based on Comparative Skeletal Anatomy: An Empirical Search for Interconnected Patterns of Natural History." *Society of Vertebrate Paleontology Memoir* 4:1–690.

Long, J. A. 1995. *The Rise of Fishes: 500 Million Years of Evolution*. Baltimore: Johns Hopkins University Press.

López-Arbarello, A., O. W. M. Rauhut, and K. Moser. 2008. "Jurassic Fishes of Gondwana." *Revista de la Asociación Geológica Argentina* 63:586–612.

Chapter 4: Smooth Amphibians

Carroll, R. L. 2007. "The Palaeozoic Ancestry of Salamanders, Frogs and Caecilians." *Zoological Journal of the Linnean Society* 150:1–140.

Holman, J. A. 2003. *Fossil Frogs and Toads of North America*. Bloomington: Indiana University Press.

———. 2006. *Fossil Salamanders of North America*. Bloomington: Indiana University Press.

Chapter 5: Mysteries Dark and Vast

Gaffney, E. S. 1975. "A Phylogeny and Classification of the Higher Categories of Turtles." *Bulletin of the American Museum of Natural History* 155 (5): 387–436.

Hay, O. P. 1908. "The Fossil Turtles of North America." *Carnegie Institution of Washington Publication* 75:586.

Joyce, W. G. 2007. "Phylogenetic Relationships of Mesozoic Turtles." *Bulletin of the Peabody Museum of Natural History* 48 (1): 3–102.

———. 2017. "A Review of the Fossil Record of Basal Mesozoic Turtles." *Bulletin of the Peabody Museum of Natural History* 58:65–113.

Lyson, T. R., G. S. Bever, B.-A. S. Bhullar, W. G. Joyce, and J. A. Gauthier. 2010. "Transitional Fossils and the Origin of Turtles." *Biology Letters* 6:830–33.

Schoch, R. R., and H.-D. Sues. 2020. "The Origin of the Turtle Body Plan: Evidence from Fossils and Embryos." *Palaeontology* 63:375–93.

Chapter 6: Beak-Heads: Ancestry of the Tuatara

Carroll, R. L., and R. Wild. 1994. "Marine Members of the Sphenodontia." In *In the Shadow of the Dinosaurs: Early Mesozoic Tetrapods*, edited by N. C. Fraser and H.-D. Sues, 70–83. New York: Cambridge University Press.

Evans, S. E., and M. E. H. Jones. 2010. "The Origin, Early History and Diversification of Lepidosauromorph Reptiles." In *New Aspects of Mesozoic Biodiversity*, edited by S. Bandyopadhyay, 27–44. Berlin: Springer.

Wu, X.-C. 1994. "Late Triassic–Early Jurassic Sphenodontians from China and the Phylogeny of the Sphenodontia." In *In the Shadow of the Dinosaurs: Early Mesozoic Tetrapods*, edited by N. C. Fraser and H.-D. Sues, 38–69. New York: Cambridge University Press.

Chapter 7: Celebration of the Lizard (and Snake)

Gilmore, C. W. 1928. "Fossil Lizards of North America." *National Academy of Sciences, Third Memoir* 22:1–201.

Keqin, G., and M. A. Norell. 2000. "Taxonomic Composition and Systematics of Late Cretaceous Lizard Assemblages from Ukhaa Tolgod and Adjacent Localities, Mongolian Gobi Desert." *Bulletin of the American Museum of Natural History* 249:1–118.

Pianka, E. R., and L. J. Vitt. 2003. *Lizards: Windows to the Evolution of Diversity*. Berkeley: University of California Press.

Rieppel, O. 1994. "The Lepidosauromorpha: An Overview with Special Emphasis on the Squamata." In *In the Shadow of the Dinosaurs: Early Mesozoic Tetrapods*, edited by N. C. Fraser and H.-D. Sues, 23–37. New York: Cambridge University Press.

Chapter 8: In the Realm of Poseidons

Everhart, M. J. 2017. *Oceans of Kansas: A Natural History of the Western Interior Sea*. 2nd ed. Bloomington: Indiana University Press.

Maisch, M. W. 2010. "Phylogeny, Systematics, and Origin of the Ichthyosauria—The State of the Art." *Palaeodiversity* 3:151–214.

Maisch, M. W., and A. T. Matzke. 2000. "The Ichthyosauria." *Stuttgarter Beiträge zur Naturkunde, Serie B (Geologie und Paläontologie)* 298:1–159.

Motani, R. 2005. "Evolution of Fish-Shaped Reptiles (Reptilia: Ichthyopterygia) in the Physical Environments and Constraints." *Annual Review of Earth and Planetary Sciences* 33:395–420.

O'Keefe, F. R. 2001. "A Cladistic Analysis and Taxonomic Revision of the Plesiosauria (Reptilia: Sauropterygia)." *Acta Zoologica Fennica* 213:1–63.

Pierce, P. 2006. *Jurassic Mary: Mary Anning and the Primeval Monsters*. Gloucestershire: History Press.

Scales, H. 2021. *The Brilliant Abyss: Exploring the Majestic Hidden Life of the Deep Ocean and the Looming Threat That Imperils It*. New York: Atlantic Monthly.

Chapter 9: Age of the Comb Jaws

Gilmore, C. W. 1928. "Fossil Lizards of North America." *National Academy of Sciences, Third Memoir* 22:1–201.

Matsumoto, R., and S. E. Evans. 2010. "Choristoderes and the Freshwater Assemblages of Laurasia." *Journal of Iberian Geology* 36:253–74.

———. 2016. "Morphology and Function of the Palatal Dentition in Choristodera." *Journal of Anatomy* 228:414–29.

Chapter 10: Something Shocking

Clark, J. M. 1994. "Patterns of Evolution in Mesozoic Crocodyliformes." In *In the Shadow of the Dinosaurs: Early Mesozoic Tetrapods*, edited by N. C. Fraser and H.-D. Sues, 84–97. New York: Cambridge University Press.

Irmis, R. B., S. J. Nesbitt, and H.-D. Sues. 2013. "Early Crocodylomorpha." *Geological Society, London, Special Publications* 379:275–302.

Schwimmer, D. R. 2002. *King of the Crocodylians: The Paleobiology of* Deinosuchus. Bloomington: Indiana University Press.

Wilberg, E. W., A. H. Turner, and C. A. Brochu. 2019. "Evolutionary Structure and Timing of Major Habitat Shifts in Crocodylomorpha." *Scientific Reports* 9:514.

Chapter 11: Wing Fingers

Dawkins, R. 1996. *Climbing Mount Improbable*. New York: W. W. Norton.

Wellnhofer, P. 1991. *The Illustrated Encyclopedia of Pterosaurs*. New York: Crescent Books.

Witton, M. P. 2013. *Pterosaurs*. Princeton, NJ: Princeton University Press.

Chapter 12: Feathered Dinosaurian Friends

Brusatte, S. 2018. *The Rise and Fall of the Dinosaurs: A New History of Their Lost World*. New York: William Morrow-Harper Collins.

Chiappe, L. M. 2007. *Glorified Dinosaurs: The Origin and Early Evolution of Birds*. Sydney: University of New South Wales Press.

Chiappe, L. M., and Q. Meng. 2016. *Birds of Stone: Chinese Avian Fossils from the Age of Dinosaurs*. Baltimore: Johns Hopkins University Press.

Feduccia, A. 1999. *The Origin and Evolution of Birds*. 2nd ed. New Haven, CT: Yale University Press.

Shipman, P. 1998. *Taking Wing:* Archaeopteryx *and the Evolution of Flight*. New York: Simon & Schuster.

Chapter 13: Copycatted

Conway Morris, S. C. 2003. *Life's Solution: Inevitable Humans in a Lonely Universe*. New York: Cambridge University Press.

Kielan-Jaworowska, Z. 2013. *In Pursuit of Early Mammals*. Bloomington: Indiana University Press.

Kielan-Jaworowska, Z., R. L. Cifelli, and Z.-X. Luo. 2004. *Mammals from the Age of Dinosaurs—Origins, Evolution, and Structure*. New York: Columbia University Press.

Lillegraven, J. A., Z. Kielan-Jaworowska, and W. A. Clemens. 1979. *Mesozoic Mammals: The First Two-Thirds of Mammalian History*. Berkeley: University of California Press.

Luo, Z.-X., Z. Kielan-Jaworowska, and R. L. Cifelli. 2002. "In Quest for a Phylogeny of Mesozoic Mammals." *Acta Palaeontologica Polonica* 47:1–78.

Rowe, T. 1988. "Definition, Diagnosis, and Origin of Mammalia." *Journal of Vertebrate Paleontology* 8:241–64.

Simpson, G. G. 1928. *A Catalogue of the Mesozoic Mammalia in the Geological Department of the British Museum*. London: British Museum (Natural History).

———. 1929. *American Mesozoic Mammalia*. London: Yale University Press.

Williamson, T. E., S. L. Brusatte, and G. P. Wilson. 2014. "The Origin and Early Evolution of Metatherian Mammals: The Cretaceous Record." *ZooKeys* 465:1–76.

Chapter 14: Epilogue

Archibald, J. D. 1996. *Dinosaurs and the End of an Era: What the Fossils Say*. New York: Columbia University Press.

Benton, M. J. 2019. *Dinosaurs Rediscovered: The Scientific Revolution in Paleontology*. New York: Thames & Hudson.

Hartman, J. H., K. R. Johnson, and D. J. Nichols. 2002. "The Hell Creek Formation and the Cretaceous-Tertiary Boundary in the Northern Great Plains: An Integrated Continental Record of the End of the Cretaceous." Geological Society of America Special Paper 361. Boulder, CO: Geological Society of America.

Ward, P. D. 2007. *Under a Green Sky: Global Warming, the Mass Extinctions of the Past, and What They Can Tell Us about Our Future*. New York: Harper Perennial.

Wilson, G. P., W. A. Clemens, J. R. Horner, and J. H. Hartman. 2014. "Through the End of the Cretaceous in the Type Locality of the Hell Creek Formation in Montana and Adjacent Areas." *Geological Society of America Special Publication* 503:392.

Index

Page numbers marked in *italics* indicate subject occurrences in figures and/or figure captions, tables, or notes.

JOHN FOSTER is a paleontologist at the Utah Field House of Natural History State Park Museum in Vernal, Utah. He has worked in the Mesozoic rocks of more than two dozen stratigraphic formations throughout the western United States for more than thirty years, specializing in the paleoecology of Late Jurassic vertebrates, including dinosaurs and other animals. Although he likes that many of the paleontological sites in the Rocky Mountain and Southwest regions are in gorgeous and wild settings, he would not mind if they were closer to the beach and good waves.

He is also author of *Jurassic West: The Dinosaurs of the Morrison Formation and Their World* (IUP, 2020) and *Cambrian Ocean World: Ancient Sea Life of North America* (IUP, 2014).

For Indiana University Press

Tony Brewer *Artist and Book Designer*

Brian Carroll *Rights Manager*

Gary Dunham *Acquisitions Editor and Director*

Anna Francis *Assistant Acquisitions Editor*

Emma Getz *Editorial Assistant*

Brenna Hosman *Production Coordinator*

Katie Huggins *Production Manager*

David Miller *Lead Project Manager/Editor*

Dan Pyle *Online Publishing Manager*

Pamela Rude *Senior Artist and Book Designer*

Stephen Williams *Marketing and Publicity Manager*